Just Enough R!

Just Enough R!

Just Enough R!
An Interactive Approach to Machine Learning and Analytics

Richard J. Roiger

Professor Emeritus, Computer Information Science Department,
Minnesota State University, Mankato

CRC Press
Taylor & Francis Group
Boca Raton London New York

CRC Press is an imprint of the
Taylor & Francis Group, an **informa** business

A CHAPMAN & HALL BOOK

First edition published 2020
by CRC Press
6000 Broken Sound Parkway NW, Suite 300, Boca Raton, FL 33487-2742

and by CRC Press
2 Park Square, Milton Park, Abingdon, Oxon, OX14 4RN

© 2020 Taylor & Francis Group, LLC

CRC Press is an imprint of Taylor & Francis Group, LLC

Library of Congress Cataloging-in-Publication Data

Names: Roiger, Richard J., author.
Title: Just enough R! : an interactive approach to machine learning and
analytics / Richard J. Roiger, Professor Emeritus, Computer Information
Science Department, Minnesota State University, Mankato.
Description: Boca Raton : CRC Press, 2020. | Includes bibliographical
references and index.
Identifiers: LCCN 2020009226 | ISBN 9780367439149 (paperback) |
ISBN 9780367443207 (hardback) | ISBN 9781003006695 (ebook)
Subjects: LCSH: R (Computer program language) | Machine learning. |
Data structures (Computer science) | Mathematical statistics—Data processing.
Classification: LCC QA276.45.R3 R65 2020 | DDC 005.13/3—dc23
LC record available at https://lccn.loc.gov/2020009226

ISBN: 978-0-367-44320-7 (hbk)
ISBN: 978-0-367-43914-9 (pbk)
ISBN: 978-1-003-00669-5 (ebk)

Typeset in Minion
by codeMantra

Contents

Preface

THE TITLE OF MY book reflects its purpose, which is to present just enough of the R language, machine learning algorithms, statistical methodology, and analytics for you to learn how to use data to solve complex problems. I wrote this book with two main goals in mind:

- To make clear how, why, and when to use machine learning techniques

- To provide what you need to become productive with R as quickly as possible

The approach is straightforward and might be called "seeing then doing, as it

- Gives step-by-step explanations using simple, understandable examples of how the various machine learning algorithms work independent of any programming language

- Explains the details of scripts—compatible with all versions of R including Version 4—that solve nontrivial problems with real data: The scripts are provided so you can watch them execute as you follow the explanations given in this book

- Implements the theory that "there's more than one way to peel an orange" by covering more than one implementation of several machine learning techniques

- Provides end-of-chapter exercises many of which can be solved by modifying an existing script

Someone once said, "In order to learn something, you must almost already know it." Several of the scripts given in this book can be thought of as problem-solving templates that, with slight modification, can be used again and again. As you develop a firm understanding of these templates, using R will become second nature.

INTENDED AUDIENCE

In writing this text, I directed my attention toward four groups of individuals:

- *Students* who want to learn about machine learning and desire hands-on experience with the R language

- *Educators* in the areas of decision science, computer science, information systems, and information technology who wish to teach a unit, workshop, or entire course on machine learning and analytics with R

- *Business professionals* who need to understand how machine learning can be applied to help solve their business problems

- *Applied researchers* who wish to add machine learning methods to their problem-solving and analytics toolkit

HOW TO USE THIS BOOK

The best way to learn quickly is to learn by both seeing and doing. We provide you this opportunity by walking you through over 50 scripts written in R. To get the most out of this book, you should first read and work through the scripts given in Chapters 1 through 4. These chapters lay the foundational work for machine learning with R.

Chapter 5 will take some time as it offers a wealth of information, some of which is statistical in nature. Here you will study linear and logistic regression as well as the Naïve Bayes classifier. You first learn the details about model evaluation using a training and test set scenario as well as cross validation. In your study of logistic regression, you will learn how to create a confusion matrix and how to create and interpret the area under a receiver operating characteristics (ROC) curve.

Once you have studied Chapter 5, Chapters 6 through 11 can be studied in any order. The only exception is that Chapter 7 should follow Chapter 6. Chapter 12 should be covered last as it offers a case study that provides insights into the entire knowledge discovery process.

SUPPLEMENTARY MATERIALS

All datasets and scripts used for illustrations and end-of-chapter exercises are described in the text. The datasets come from several areas including business, health and medicine, and science. The datasets and scripts can be downloaded at two locations:

- The CRC Web site: https://www.crcpress.com/9780367439149

- https://krypton.mnsu.edu/~sa7379bt/

If you have trouble obtaining any of these materials, please email me at: richard.roiger@mnsu.edu

INSTRUCTOR SUPPLEMENTS

Supplements are provided to help the instructor organize lectures and write examinations.

Please note that these supplements are available to qualified instructors only. Contact your CRC sales representative or get help by visiting https://www.crcpress.com/contactus to access this material.

Acknowledgment

I AM DEEPLY INDEBTED TO my wife, Suzanne, for her extreme patience and consistent support.

Author

Richard J. Roiger is a professor emeritus at Minnesota State University, Mankato, where he taught and performed research in the Computer and Information Science Department for more than 30 years. Dr. Roiger's PhD degree is in Computer and Information Sciences from the University of Minnesota. He has presented many conference papers and written several journal articles about topics in machine learning and knowledge discovery. After his retirement, Dr. Roiger continues to serve as a part-time faculty member teaching courses in machine learning, artificial intelligence, and research methods. He is a board member of the Retired Education Association of Minnesota where he serves as the financial advisor. He enjoys interacting with his grandchildren, traveling, writing, and pursuing his musical talents.

Introduction to Machine Learning

In This Chapter

- Definitions and Terminology

- Machine Learning Strategies

- Evaluation Techniques

- Ethical Issues

THE R LANGUAGE CONTINUES to maintain its status as one of the top-rated problem-solving tools within the areas of machine learning, data science, data analytics, data mining, and statistical analysis. It's easy to see why: R is free, contains thousands of packages, is supported by a growing community of users, and is easy to use when interfaced with RStudio's integrated development environment!

R's popularity has resulted in the development of thousands of tutorials on machine learning. The information is all there! Unfortunately, it's easy to get lost in a maze of too much information. Valuable time is spent trying to find exactly what is needed to solve problems. The end result is frustration and difficulty understanding what's important.

We believe our approach of presenting and clearly explaining script-based problem-solving techniques provides the tools you need for machine learning with R. The book's title reflects its purpose. *Just Enough R!* gives you just enough of the R language and machine learning methods to minimize stumbling blocks and cut through the maze. Our goal is to give you what you need to become productive with R as quickly as possible.

In this chapter, we offer a brief introduction to machine learning. In Chapter 2, we move right into the nuts and bolts of the *R* language and the problem-solving techniques it offers. We conclude this chapter with a short summary, key term definitions, and a set of exercises. Let's get started!

1.1 MACHINE LEARNING, STATISTICAL ANALYSIS, AND DATA SCIENCE

It's almost impossible to surf the Web, open a newspaper, or turn on the TV without being exposed to terms such as machine learning, statistical analysis, data science, data analytics, and data mining. Most people have some idea about what these terms mean, but if you ask for a precise definition of any of them, you get a variety of answers. Here are a few distinctions:

- Building models to find structure in data has its roots in the fields of mathematics and statistics. *Statistical methods* are differentiated from other techniques in that they make certain assumptions about the nature of the data. Technically, if these assumptions are violated, the models built with these techniques may be inaccurate.

- *Machine learning* can be differentiated from statistical modeling in that assumptions about data distributions and variable independence are not a concern. Machine learning is generally considered an area of specialization within the broader field of artificial intelligence. However, most textbooks make little or no distinction between machine learning and statistical methods.

- *Data science* or *data analytics* is often defined as the process of extracting meaningful knowledge from data. Its methods come from several disciplines including computer science, mathematics, statistics, data warehousing, and distributed processing to name a few. Although machine learning is often seen in data science applications, it is not required.

- *Data mining* first became popular in the academic community about 1995 and can be defined as the process of using one or several machine learning algorithms to find structure in data. The structure may take many forms including a set of rules, a graph or network, a tree, one or several equations, and more. The structure can be part of a complex visual dashboard or as simple as a list of political candidates and an associated number representing voter sentiment based on twitter feeds.

- The phrase *knowledge discovery in databases* (KDD) was coined in 1989 to emphasize that knowledge can be derived from data-driven discovery and is frequently used interchangeably with data mining. In addition to performing data mining, a typical KDD process model includes a methodology for extracting and preparing data as well as making decisions about actions to be taken once data mining has taken place. As much of today's data is not found in a traditional data warehouse, KDD is most often associated with *knowledge discovery in data*.

Although these general distinctions might be made, the most important point is that all of these terms define techniques designed to solve problems by finding interesting structure in data. We prefer to use the term *machine learning* as our focus is both on how to apply the algorithms and on understanding how the algorithms work. However, we often interchange the terms machine learning and data mining.

1.2 MACHINE LEARNING: A FIRST EXAMPLE

Supervised learning is probably the best and most widely used technique for machine learning. The purpose of supervised learning is twofold. First, we use supervised learning to build classification models from sets of data containing examples and nonexamples of the concepts to be learned. Each example or nonexample is known as an *instance* of data. Second, once a classification model has been constructed, the model is used to determine the classification of newly presented instances of unknown origin. It is worth noting that although model creation is inductive, applying the model to classify new instances of unknown origin is a deductive process.

1.2.1 Attribute-Value Format

To more clearly illustrate the idea of supervised learning, consider the hypothetical dataset shown in Table 1.1. The dataset is very small and is relevant for illustrative purposes only. The table data is displayed in *attribute-value format* where the first row shows names for the attributes whose values are contained in the table. The attributes *sore throat, fever, swollen glands, congestion,* and *headache* are possible symptoms experienced by individuals who have a particular affliction (a *strep throat,* a *cold,* or an *allergy*). These attributes are known as *input attributes* and are used to create a model to represent the data. *Diagnosis* is the attribute whose value we wish to predict. *Diagnosis* is known as the *class, response,* or *output attribute*.

Starting with the second row of the table, each remaining row is an *instance* of data. An individual row shows the symptoms and affliction of a single patient. For example, the patient with ID = 1 has a sore throat, fever, swollen glands, congestion, and a headache. The patient has been diagnosed as having strep throat.

Suppose we wish to develop a generalized model to represent the data shown in Table 1.1. Even though this dataset is small, it would be difficult for us to develop a general representation unless we knew something about the relative importance of the individual attributes and possible relationships among the attributes. Fortunately, an appropriate supervised learning algorithm can do the work for us.

TABLE 1.1 Hypothetical Training Data for Disease Diagnosis

Patient ID#	Sore Throat	Fever	Swollen Glands	Congestion	Headache	Diagnosis
1	Yes	Yes	Yes	Yes	Yes	Strep throat
2	No	No	No	Yes	Yes	Allergy
3	Yes	Yes	No	Yes	No	Cold
4	Yes	No	Yes	No	No	Strep throat
5	No	Yes	No	Yes	No	Cold
6	No	No	No	Yes	No	Allergy
7	No	No	Yes	No	No	Strep throat
8	Yes	No	No	Yes	Yes	Allergy
9	No	Yes	No	Yes	Yes	Cold
10	Yes	Yes	No	Yes	Yes	Cold

1.2.2 A Decision Tree for Diagnosing Illness

We presented the data in Table 1.1 to C4.5 (Quinlan, 1993), a supervised learning program that generalizes a set of input instances by building a decision tree. A *decision tree* is a simple structure where nonterminal nodes represent tests on one or more attributes and terminal nodes reflect decision outcomes. Decision trees have several advantages in that they are easy for us to understand, can be transformed into rules, and have been shown to work well experimentally. A supervised algorithm for creating a decision tree is detailed in Chapter 6.

Figure 1.1 shows the decision tree created from the data in Table 1.1. The decision tree generalizes the table data. Specifically,

- If a patient has swollen glands, the disease diagnosis is strep throat.

- If a patient does not have swollen glands and has a fever, the diagnosis is cold.

- If a patient does not have swollen glands and does not have a fever, the diagnosis is allergy.

The decision tree tells us that we can accurately diagnose a patient in this dataset by concerning ourselves only with whether the patient has swollen glands and a fever. The attributes sore throat, congestion, and headache do not play a role in determining a diagnosis. As you can see, the decision tree has generalized the data and provided us with a summary of those attributes and attribute relationships important for an accurate diagnosis.

Let's use the decision tree to classify the first two instances shown in Table 1.2.

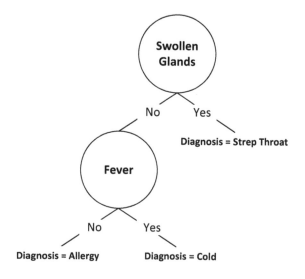

FIGURE 1.1   A decision tree for the data in Table 1.1.

TABLE 1.2   Data Instances with an Unknown Classification

Patient ID	Sore Throat	Fever	Swollen Glands	Congestion	Headache	Diagnosis
11	No	No	Yes	Yes	Yes	?
12	Yes	Yes	No	No	Yes	?
13	No	No	No	No	Yes	?

- Since the patient with ID = 11 has a value of *Yes* for *swollen glands*, we follow the right link from the root node of the decision tree. The right link leads to a terminal node, indicating the patient has strep throat.

- The patient with ID = 12 has a value of *No* for *swollen glands*. We follow the left link and check the value of the attribute *fever*. Since *fever* equals *Yes*, we diagnose the patient with a cold.

We can translate any decision tree into a set of *production rules*. Production rules are rules of the form:

IF *antecedent conditions*

THEN *consequent conditions*

The antecedent conditions detail values or value ranges for one or more input attributes. The consequent conditions specify the values or value ranges for the output attributes. The technique for mapping a decision tree to a set of production rules is simple. A rule is created by starting at the root node and following one path of the tree to a leaf node. The antecedent of a rule is given by the attribute-value combinations seen along the path. The consequent of the corresponding rule is the value at the leaf node. Here are the three production rules for the decision tree shown in Figure 1.1:

1. IF *Swollen Glands = Yes*
   THEN *Diagnosis = Strep Throat*

2. IF *Swollen Glands = No & Fever = Yes*
   THEN *Diagnosis = Cold*

3. IF *Swollen Glands = No & Fever = No*
   THEN *Diagnosis = Allergy*

Let's use the production rules to classify the table instance with patient ID = 13. Because *swollen glands* equals *No*, we pass over the first rule. Likewise, because *fever* equals *No*, the second rule does not apply. Finally, both antecedent conditions for the third rule are satisfied. Therefore, we are able to apply the third rule and diagnose the patient as having an allergy.

The instances used to create the decision tree model are known as *training data*. At this point, the training instances are the only instances known to be correctly classified by the model. However, our model is useful to the extent that it can correctly classify new instances whose classification is not known. To determine how well the model is able to be of general use, we test the accuracy of the model using a *test set*. The instances of the test set have a known classification. Therefore, we can compare the test set instance classifications determined by the model with the correct classification values. Test set classification correctness gives us some indication about the future performance of the model.

Our decision tree example is a very simple example of one of several supervised learning techniques. You will learn much more about decision trees in Chapter 6. Other supervised techniques include rule-based systems (Chapter 7), some neural network structures (Chapter 8), linear regression (Chapter 5), and many more. The next section examines the general categories of machine learning strategies.

1.3 MACHINE LEARNING STRATEGIES

Machine learning *strategies* can be broadly classified as either supervised or unsupervised. Supervised learning builds models by using input attributes to predict output attribute values. Many supervised learning algorithms only permit a single output attribute. Other supervised learning tools allow us to specify one or several output attributes. Output attributes are also known as *dependent variables* as their outcome depends on the values of one or more input attributes. Input attributes are referred to as *independent variables*. When learning is unsupervised, an output attribute does not exist. Therefore, all attributes used for model building are independent variables.

Supervised learning strategies can be further labeled according to whether output attributes are discrete or categorical, as well as by whether the models are designed to determine a current condition or predict future outcome. In this section, we examine three supervised learning strategies, take a closer look at unsupervised clustering, and introduce a strategy for discovering associations among retail items sold in catalogs and stores. Figure 1.2 shows the five machine learning strategies we will discuss.

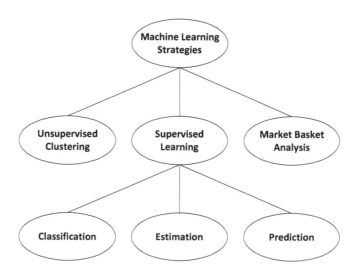

FIGURE 1.2   A hierarchy of machine learning strategies.

1.3.1 Classification

Classification is probably the best understood of all strategies. Classification tasks have three common characteristics:

- Learning is supervised.

- The dependent variable is categorical.

- The emphasis is on building models able to assign new instances to one of a set of well-defined classes.

Some example classification tasks include the following:

- Determine those characteristics that differentiate individuals who have suffered a heart attack from those who have not.

- Develop a profile of a "successful" person.

- Determine if a credit card purchase is fraudulent.

- Classify a car loan applicant as a good or a poor credit risk.

- Determine if an income tax submission has been falsified.

Notice that each example deals with current, rather than future, behavior. For example, we want the car loan application model to determine whether an applicant is a good credit risk at this time rather than in some future time period. Prediction models are designed to answer questions about future behavior. Prediction models are discussed later in this section.

1.3.2 Estimation

Like classification, the purpose of an *estimation* model is to determine a value for an unknown output attribute. However, unlike classification, the output attribute for an estimation problem is numeric rather than categorical. Here are three examples of estimation tasks:

- Estimate the number of minutes before a thunderstorm will reach a given location.

- Estimate the return on investment for an advertising campaign.

- Estimate the likelihood that a credit card has been stolen.

Most supervised techniques are able to solve classification or estimation problems, but not both. If our learning tool supports one strategy but not the other, we can usually adapt a problem for solution by either strategy. To illustrate, suppose the output attribute for the original training data in the stolen credit card example above is a numeric value between 0 and 1, with 1 being a most likely case for a stolen card. We can make discrete categories for the output attribute values by replacing scores ranging between 0.0 and 0.3 with the value *unlikely*, scores between 0.3 and 0.7 with *likely*, and scores greater than 0.7 with *highly likely*. In this case, the transformation between numeric values and discrete categories is straightforward. Cases such as attempting to make monetary amounts discrete present more of a challenge.

THE CARDIOLOGY PATIENT DATASET

The original cardiology patient data was gathered by Dr. Robert Detrano at the VA Medical Center in Long Beach, California. The dataset consists of 303 instances. One hundred thirty-eight of the instances hold information about patients with heart disease. The original dataset contains 13 numeric attributes and a 14th attribute indicating whether the patient has a heart condition. The dataset was later modified by Dr. John Gennari. He changed seven of the numerical attributes to categorical equivalents for the purpose of testing machine learning tools able to classify datasets with mixed data types. The files are named CardiologyNumerical and CardiologyMixed, respectively. This dataset is interesting because it represents real patient data and has been used extensively for testing various machine learning techniques. We can use this data together with one or more machine learning techniques to help us develop profiles for differentiating individuals with heart disease from those without known heart conditions.

1.3.3 Prediction

It is not easy to differentiate prediction from classification or estimation. However, unlike a classification or estimation model, the purpose of a predictive model is to determine future outcome rather than current behavior. The output attribute(s) of a predictive model can be categorical or numeric. Here are several examples of tasks appropriate for predictive data mining:

- Predict the total number of touchdowns a National Football League (NFL) running back will score during the current NFL season.

- Determine whether a credit card customer is likely to take advantage of a special offer made available with their credit card billing.

- Predict next week's closing price for the Dow Jones Industrial Average.

- Forecast which cell phone subscribers are likely to change providers during the next 3 months.

Most supervised machine learning techniques appropriate for classification or estimation problems can also build predictive models. Actually, it is the nature of the data that determines whether a model is suitable for classification, estimation, or prediction. To show this, let's consider a real medical dataset with 303 instances. One hundred sixty-five instances hold information about patients who are free of heart disease. The remaining 138 instances contain data about patients who have a known heart condition.

The attributes and possible attribute values associated with this dataset are shown in Table 1.3. Two forms of the dataset exist. One dataset consists of all numeric attributes.

TABLE 1.3   Cardiology Patient Data

Attribute Name	Mixed Values	Numeric Values	Comments
Age	Numeric	Numeric	Age in years
Gender	Male, Female	1, 0	Patient gender
Chest pain type	Angina, Abnormal Angina, NoTang, Asymptomatic	1–4	NoTang = Nonanginal pain
Blood pressure	Numeric	Numeric	Resting blood pressure upon hospital admission
Cholesterol	Numeric	Numeric	Serum cholesterol
Fasting blood sugar < 120	True, False	1, 0	Is fasting blood sugar less than 120?
Resting CG	Normal, Abnormal, Hyp	0, 1, 2	Hyp = Left ventricular hypertrophy
Maximum heart rate	Numeric	Numeric	Maximum heart rate achieved
Induced angina?	True, False	1, 0	Does the patient experience angina as a result of exercise?
Old peak	Numeric	Numeric	ST depression induced by exercise relative to rest
Slope	Up, Flat, Down	1–3	Slope of the peak exercise ST segment
Number of Colored Vessels	0, 1, 2, 3	0, 1, 2, 3	Number of major vessels colored by fluoroscopy
Thal	Normal fix, rev	3, 6, 7	Normal, fixed defect, reversible defect
Concept class	Healthy, Sick	1, 0	Angiographic disease status

The second dataset has categorical conversions for seven of the original numeric attributes. The table column labeled *Mixed Values* shows the value *Numeric* for attributes that were not converted to a categorical equivalent. For example, the values for attribute *Age* are identical for both datasets. However, the attribute *Fasting Blood Sugar < 120* has values *True* or *False* in the converted dataset and values 0 and 1 in the original data.

Table 1.4 lists four instances from the mixed form of the dataset. Two of the instances represent typical examples from each respective class. The remaining two instances are atypical class members. Some differences between a typical healthy and a typical sick patient are easily anticipated, for example, *Resting ECG* and *Induced Angina*. Surprisingly, we do not see expected differences in cholesterol and blood pressure readings between most and least typical healthy and individuals.

Here are two rules generated from this dataset. *Class* is specified as the output attribute:

- IF *Maximum Heart Rate >= 158.333*
  THEN *Class = Healthy*
      Rule Precision: 75.60%
      Rule Coverage: 40.60%

TABLE 1.4   Typical and Atypical Instances from the Cardiology Domain

Attribute Name	Typical Healthy Class	Atypical Healthy Class	Typical Sick Class	Atypical Sick Class
Age	52	63	60	62
Gender	Male	Male	Male	Female
Chest pain type	NoTang	Angina	Asymptomatic	Asymptomatic
Blood pressure	138	145	125	160
Cholesterol	223	233	258	164
Fasting blood sugar < 120	False	True	False	False
Resting ECG	Normal	Hyp	Hyp	Hyp
Maximum heart rate	169	150	141	145
Induced angina?	False	False	True	False
Old peak	0	2.3	2.8	6.2
Slope	Up	Down	Flat	Down
Number of colored vessels	0	0	1	3
Thal	Normal	Fix	Rev	Rev

- IF *Thal* = *Rev*
  THEN *Class* = *Sick*
  > Rule Precision: 76.30%
  > Rule Coverage: 38.90%

Let's consider the first rule. Rule coverage gives us the percent of data instances that satisfy the rule's preconditions. Rule coverage reveals that of the 303 individuals represented in the dataset, 40.6% or 123 patients have a maximum heart rate greater than 158.333. Rule precision tells us that of the 123 individuals with maximum heart rate ≥ 158.333, 75.6% or 93 are healthy patients. That is, if a patient in the dataset has a maximum heart rate greater than 158.333, we will be correct more than 75 times out of 100 in identifying the patient as healthy. When we combine this knowledge with the maximum heart rate values shown in Table 1.4, we are able to conclude that healthy patients are likely to have higher maximum heart rate values.

Is this first rule appropriate for classification or prediction? If the rule is predictive, we can use it to warn healthy folks with the statement:

Warning 1: Have your maximum heart rate checked on a regular basis. If your maximum heart rate is low, you may be at risk of having a heart attack!

If the rule is appropriate for classification but not prediction, the scenario reads:

Warning 2: If you have a heart attack, expect your maximum heart rate to decrease.

In any case, we cannot imply the stronger statement:

Warning 3: A low maximum heart rate will cause you to have a heart attack!

That is, with machine learning, we can state relationships between attributes, but we cannot say whether the relationships imply causality. Therefore, entertaining an exercise program to increase maximum heart rate may or may not be a good idea.

The question still remains as to whether either of the first two warnings is correct. This question is not easily answered. A specialist can develop models to generate rules such as those just given. Beyond this, the specialist must have access to additional information—in this case, a medical expert—before determining how to use discovered knowledge.

1.3.4 Unsupervised Clustering

With unsupervised clustering (Chapter 11), we are without a dependent variable to guide the learning process. Rather, the learning program builds a knowledge structure by using some measure of cluster quality to group instances into two or more classes. A primary goal of an unsupervised clustering strategy is to discover concept structures in data. Common uses of unsupervised clustering include the following:

- Detect fraud in stock trading activity, insurance claims, or financial transactions

- Determine if meaningful relationships in the form of concepts can be found in data representing gamma-ray burst flashes outside of our solar system

- Determine a best set of input attributes for building a supervised model to identify cell phone customers likely to churn

It is not unusual to use unsupervised clustering as an evaluation tool for supervised learning. To illustrate this idea, let's suppose we have built a supervised model using the heart patient data with output attribute *Class*. To evaluate the supervised model, we present the training instances to an unsupervised clustering system. The attribute *Class* is flagged as unused. Next, we examine the output of the unsupervised model to determine if the instances from each concept class (*Healthy* and *Sick*) naturally cluster together. If the instances from the individual classes do not cluster together, we may conclude that the attributes are unable to distinguish healthy patients from those with a heart condition. This being the case, the supervised model is likely to perform poorly. One solution is to choose a best set of attributes for a supervised learner model by repeatedly applying unsupervised clustering with alternative attribute choices. In this way, those attributes best able to differentiate the classes known to be present in the data can be determined. Unfortunately, even with a small number of attribute choices, the application of this technique can be computationally unmanageable.

Unsupervised clustering can also help detect any atypical instances present in the data, that is, instances that do not group naturally with the other instances. Atypical instances are referred to as *outliers*. Outliers can be of great importance and should be identified whenever possible. With machine learning, the outliers might be just those instances we are trying to identify. For example, an application that checks credit card purchases would likely identify an outlier as a positive instance of credit card fraud.

1.3.5 Market Basket Analysis

The purpose of *market basket analysis* is to find interesting relationships among retail products. The results of a market basket analysis help retailers design promotions, arrange shelf or catalog items, and develop cross-marketing strategies. Association rule algorithms are often used to apply a market basket analysis to a set of data. Association rules are detailed in Chapter 7.

1.4 EVALUATING PERFORMANCE

Performance evaluation is probably the most critical of all the steps in the analytics process. In this section, we offer a common sense approach to evaluating supervised and unsupervised models. In later chapters, we will concentrate on more formal evaluation techniques. As a starting point, we pose three general questions:

1. Will the benefits received from a machine learning project more than offset cost?

2. How do we interpret the results?

3. Can we use the results with confidence?

The first question requires knowledge about the business model, the current state of available data, and current resources. Therefore, we will turn our attention to providing evaluation tools for questions 2 and 3. We first consider the evaluation of supervised models.

1.4.1 Evaluating Supervised Models

Supervised models are designed to classify, estimate, and/or predict future outcomes. For some applications, the desire is to build models showing consistently high predictive accuracy. The following three applications focus on classification correctness:

- Develop a model to accept or reject credit card applicants

- Develop a model to accept or reject home mortgage applicants

- Develop a model to decide whether or not to drill for oil

Classification correctness is best calculated by presenting previously unseen data in the form of a test set to the model being evaluated. Test set model accuracy can be summarized in a table known as a *confusion matrix*. To illustrate, let's suppose we have three possible classes: C_1, C_2, and C_3. A generic confusion matrix for the three-class case is shown in Table 1.5.

Values along the main diagonal give the total number of correct classifications for each class. For example, a value of 15 for C_{11} means that 15 class C_1 test set instances were

TABLE 1.5   A Three-Class Confusion
Matrix

Computed Decision

	C_1	C_2	C_3
C_1	C_{11}	C_{12}	C_{13}
C_2	C_{21}	C_{22}	C_{23}
C_3	C_{31}	C_{32}	C_{33}

correctly classified. Values other than those on the main diagonal represent classification errors. To illustrate, suppose C_{12} has the value 4. This means that four class C_1 instances were incorrectly classified as belonging to class C_2. The following four observations may be helpful in analyzing the information in a confusion matrix:

1. For any given cell C_{ij}, the subscript i represents the actual class and j indicates the computed class.

2. Values along the main diagonal represent correct classifications. For the matrix in Table 1.5, the value C_{11} represents the total number of class C_1 instances correctly classified by the model. A similar statement can be made for the values C_{22} and C_{33}.

3. Values in row C_i represent those instances that belong to class C_i. For example, with $i = 2$, the instances associated with cells C_{21}, C_{22}, and C_{23} are all actually members of C_2. To find the total number of C_2 instances incorrectly classified as members of another class, we compute the sum of C_{21} and C_{23}.

4. Values found in column C_i indicate those instances that have been classified as members of C_i. With $i = 2$, the instances associated with cells C_{12}, C_{22}, and C_{32} have been classified as members of class C_2. To find the total number of instances incorrectly classified as members of class C_2, we compute the sum of C_{12} and C_{32}.

We can use the summary data displayed in the confusion matrix to compute model accuracy. To determine the accuracy of a model, we sum the values found on the main diagonal and divide the sum by the total number of test set instances. For example, if we apply a model to a test set of 100 instances and the values along the main diagonal of the resultant confusion matrix sum to 70, the test set accuracy of the model is 0.70 or 70%. As model accuracy is often given as an error rate, we can compute model error rate by subtracting the model accuracy value from 1.0. For our example, the corresponding error rate is 0.30.

1.4.2 Two-Class Error Analysis

The three applications listed at the beginning of this section represent two-class problems. For example, a credit card application is either accepted or rejected. We can use a simple two-class confusion matrix to help us analyze each of these applications.

Consider the confusion matrix displayed in Table 1.6. Cells showing *True Accept* and *True Reject* represent correctly classified test set instances. For the first and second applications

TABLE 1.6   A Simple Confusion Matrix

	Computed Accept	Computed Reject
Accept	True Accept	False Reject
Reject	False Accept	True Reject

presented in the previous section, the cell with *False Accept* denotes accepted applicants that should have been rejected. The cell with *False Reject* designates rejected applicants that should have been accepted. A similar analogy can be made for the third application. Let's use the confusion matrices shown in Table 1.7 to examine the first application in more detail.

Assume the confusion matrices shown in Table 1.7 represent the test set error rates of two supervised learner models built for the credit card application problem. The confusion matrices show that each model displays an error rate of 10%. As the error rates are identical, which model is better? To answer the question, we must compare the average cost of credit card payment default to the average potential loss in profit realized by rejecting individuals who are good approval candidates. Given that credit card purchases are unsecured, the cost of accepting credit card customers likely to default is more of a concern. In this case, we should choose model B because the confusion matrices tell us that this model is less likely to erroneously offer a credit card to an individual likely to default. Does the same reasoning apply for the home mortgage application? How about the application where the question is whether to drill for oil? As you can see, although test set error rate is a useful measure for model evaluation, other factors such as costs incurred for false inclusion as well as losses resulting from false omission must be considered.

1.4.3 Evaluating Numeric Output

A confusion matrix is of little use for evaluating supervised models offering numeric output. In addition, the concept of classification correctness takes on a new meaning with numeric output models because instances cannot be directly categorized into one of several possible output classes. However, several useful measures of model accuracy have been defined for supervised models having numeric output. The most common numeric accuracy measures are mean absolute error and mean squared error.

The *mean absolute error (mae)* for a set of test data is computed by finding the average absolute difference between computed and desired outcome values. In a similar manner, the *mean squared error* is the average squared difference between computed and desired outcomes. Finally, the *root mean squared error (rms)* is simply the square root of a mean squared error value. *Rms* is frequently used as a measure of test set accuracy with

TABLE 1.7   Two Confusion Matrices Each Showing a 10% Error Rate

Model A	Computed Accept	Computed Reject	Model B	Computed Accept	Computed Reject
Accept	600	25	Accept	600	75
Reject	75	300	Reject	25	300

feed-forward neural networks. It is obvious that for a best test set accuracy, we wish to obtain the smallest possible value for each measure. You will learn more about the pros and cons of each of these numeric measures of accuracy in Chapter 9.

1.4.4 Comparing Models by Measuring Lift

Marketing applications that focus on response rates from mass mailings are less concerned with test set classification error and more interested in building models able to extract bias samples from large populations. Bias samples show higher response rates than the rates seen within the general population. Supervised models designed for extracting bias samples from a general population are often evaluated by a measure that comes directly from marketing known as *lift*. An example illustrates the idea.

Suppose the Acme Credit Card Company is about to launch a new promotional offer with next month's credit card statement. The company has determined that for a typical month, approximately 100,000 credit card holders show a zero balance on their credit card. The company has also determined that an average of 1% of all card holders take advantage of promotional offers included with their card billings. Based on this information, approximately 1000 of the 100,000 zero-balance card holders are likely to accept the new promotional offer. As zero-balance card holders do not require a monthly billing statement, the problem is to send a zero-balance billing to exactly those customers who will accept the new promotion.

We can employ the concept of lift to help us choose a best solution. Lift measures the change in percent concentration of a desired class, C_i, taken from a biased sample relative to the concentration of C_i within the entire population. We can formulate lift using conditional probabilities. Specifically,

$$Lift = \frac{P(C_i \mid Sample)}{P(C_i \mid Population)}$$

where $P(C_i \mid Sample)$ is the portion of instances contained in class C_i relative to the biased sample population and $P(C_i \mid Population)$ is the fraction of class C_i instances relative to the entire population. For our problem, C_i is the class of all zero-balance customers who, given the opportunity, will take advantage of the promotional offer.

Figure 1.3 offers a graphical representation of the credit card promotion problem. The graph is sometimes called a *lift chart*. The horizontal axis shows the percent of the total population sampled, and the vertical axis represents the number of likely respondents. The graph displays model performance as a function of sample size. The straight line represents the general population. This line tells us that if we randomly select 20% of the population for the mailing, we can expect a response from 200 of the 1000 likely respondents. Likewise, selecting 100% of the population will give us all respondents. The curved line shows the lift achieved by employing models of varying sample sizes. By examining the graph, you can see that an ideal model will show the greatest lift with the smallest sample size. This is represented in Figure 1.3 as the upper-left portion of the graph. Although Figure 1.3 is useful,

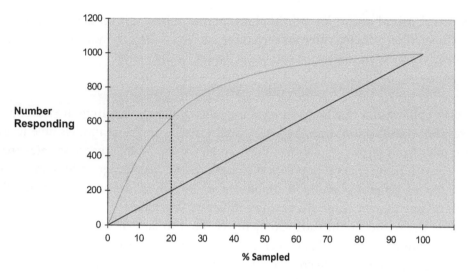

FIGURE 1.3   Targeted vs. mass mailing.

the confusion matrix also offers us an explanation about how lift can be incorporated to solve problems.

Table 1.8 shows two confusion matrices to help us understand the credit card promotion problem from the perspective of lift. The confusion matrix showing *No Model* tells us that all zero-balance customers are sent a billing statement with the promotional offer. By definition, the lift for this scenario is 1.0 because the sample and the population are identical. The lift for the matrix showing *Ideal Model* is 100 (100%/1%) because the biased sample contains only positive instances.

Consider the confusion matrices for the two models shown in Table 1.9. The lift for model X is computed as:

$$Lift(\text{model X}) = \frac{540/24000}{1000/100000},$$

TABLE 1.8   Two Confusion Matrices: No Model and an Ideal Model

No Model	Computed Accept	Computed Reject	Ideal Model	Computed Accept	Computed Reject
Accept	1,000	0	Accept	1,000	0
Reject	99,000	0	Reject	0	99,000

TABLE 1.9   Two Confusion Matrices for Alternative Models with Lift Equal to 2.25

No Model	Computed Accept	Computed Reject	Ideal Model	Computed Accept	Computed Reject
Accept	540	460	Accept	450	550
Reject	23,460	75,540	Reject	19,550	79,450

which evaluates to 2.25. The lift for model Y is computed as:

$$Lift(\text{model Y}) = \frac{450/20000}{1000/100000},$$

which also evaluates to 2.25. As was the case with the previous example, to answer the question about which is a better model, we must have additional information about the relative costs of false negative and false positive selections. For our example, model Y is a better choice if the cost savings in mailing fees (4000 fewer mailings) more than offsets the loss in profits incurred from fewer sales (90 fewer sales).

1.4.5 Unsupervised Model Evaluation

Evaluating unsupervised models is, in general, a more difficult task than supervised evaluation. This is true because the goals of an unsupervised session are frequently not as clear as the goals for supervised learning. All unsupervised clustering techniques compute some measure of cluster quality. A common technique is to calculate the summation of squared error differences between the instances of each cluster and their corresponding cluster center. Smaller values for sums of squared error differences indicate clusters of higher quality. A second approach compares the ratio of within-cluster to between-cluster similarity values. In Chapter 11, you will see a third evaluation technique where supervised learning is used to evaluate unsupervised clustering.

Finally, a common misconception in the business world is that machine learning can be accomplished simply by choosing the right tool, turning it loose on some data, and waiting for answers to problems. This approach is doomed to failure. Machines are still machines. It is the analysis of results provided by the human element that ultimately dictates the success or failure of a machine learning project. A formal process model such as the one described in Chapter 4 will help provide more complete answers to the questions posed at the beginning of this section.

1.5 ETHICAL ISSUES

Machine learning models offer sophisticated tools to deduce sensitive patterns from data. Because of this, the use of data containing information about people requires special precautions. Let's look at two examples.

Human capital management (HCM) is an approach that perceives employees as assets with measurable value. The employees of a company that invests in HCM receive consistent communication about their performance expectations. Managers objectively rate employees, thereby holding them accountable for specific goals. Employees that meet or exceed expectations are rewarded and those consistently falling short are terminated. With this type of system, it is easy to keep objective records of current employees and to maintain previous employee records in a warehouse facility.

Machine learning can be used in a positive way as one of several tools to help associate current employees with tasks or positions for which they are best suited. It can also be misused by building a classification model that differentiates former employees relative to their termination. Possibilities include that a former employee quit, was fired, was laid off, retired, or is deceased. The employer then applies the model built from former employee data to current employees and fires those the model predicts are likely to quit. Furthermore, the employer may use the model to lay off or fire employees predicted to retire soon. This use of machine learning is certainly less than ethical.

As a second example, anonymized health records are public information as are people's names. However, being able to associate people with their individual health records results in creating private information. This process of being able to deduce unauthorized or private information from publicly available data is known as the *inference problem* (Grossman et al., 2002). Here are three ways to approach this issue.

- Given a database and a machine learning tool, apply the tool to determine if sensitive information can be deduced

- Use an inference controller to detect the motives of the user

- Give limited samples of the data to the user, thereby preventing the user from building a machine learning model

Each approach is appropriate in certain situations but not others. When privacy of the individual is at stake, exercising wisdom and great caution is the best approach.

1.6 CHAPTER SUMMARY

Machine learning is used by thousands of companies and organizations to monitor user experiences, calculate return on investment for advertising campaigns, predict flooding, detect fraud, and even help design vehicles. The applications seem almost endless.

A model created by a machine learning algorithm is a conceptual generalization of the data. Common generalizations take the form of a tree, a network, one or more equations, or a set of rules. With supervised learning, instances whose classification is known are used by the machine learning tool to build a general model representing the data. The created model is then employed to determine the classification of new, previously unclassified instances. With unsupervised clustering, predefined concepts do not exist. Instead, data instances are grouped together based on a similarity scheme defined by the clustering model.

In the next chapters, you will learn more about the steps of the machine learning process. You will learn about different machine learning algorithms, as well as several techniques to help you determine when to apply machine learning to your problems. A common theme we hope to convey throughout this book is that machine learning is about model building and not about magic. Human nature requires that we generalize and categorize the world around us. For this reason, model building is a natural process that can be fun and very rewarding!

1.7 KEY TERMS

- *Attribute-value format.* A table format where the first row of the table contains an attribute name. Each row of the table after the first contains a data instance whose attribute values are given in the columns of the table.

- *Classification.* A supervised learning strategy where the output attribute is categorical. Emphasis is on building models able to assign new instances to one of a set of well-defined classes.

- *Confusion matrix.* A matrix used to summarize the results of a supervised classification. Entries along the main diagonal represent the total number of correct classifications. Entries other than those on the main diagonal represent classification errors.

- *Data warehouse.* A historical database designed for decision support rather than transaction processing.

- *Decision tree.* A tree structure where nonterminal nodes represent tests on one or more attributes and terminal nodes reflect decision outcomes.

- *Dependent variable.* A variable whose value is determined by one or more independent variables.

- *Estimation.* A supervised learning strategy where the output attribute is numeric. Emphasis is on determining current, rather than future, outcome.

- *Independent variable.* Input attributes used for building supervised or unsupervised learner models.

- *Input attribute.* An attribute used by a machine learning algorithm to help create a model of the data. An input attribute is sometimes referred to as an independent variable.

- *Instance.* An example or nonexample of a concept.

- *Lift.* The probability of class C_i given a sample taken from population P divided by the probability of C_i given the entire population P.

- *Lift chart.* A graph that displays the performance of a machine learning model as a function of sample size.

- *Market basket analysis.* A machine learning strategy that attempts to find interesting relationships among retail products.

- *Mean absolute error.* For a set of training or test set instances, the mean absolute error is the average absolute difference between classifier-predicted output and actual output.

- *Mean squared error.* For a set of training or test set instances, the mean squared error is the average of the sum of squared differences between classifier-predicted output and actual output.

- *Outlier.* An atypical data instance.

- *Output attribute.* For supervised learning, the attribute whose output is to be predicted.

- *Prediction.* A supervised learning strategy designed to determine future outcome.

- *Production rule.* A rule of the form: IF antecedent conditions THEN consequent conditions.

- *Root mean squared error.* The square root of the mean squared error.

- *Rule coverage.* Given rule R, rule coverage is the percent of all instances that satisfy R's preconditions.

- *Rule precision.* For rule R, the percent of instances covered by R's antecedent condition(s) that are also covered by R's consequent. Rule precision and rule accuracy are often assumed to be interchangeable terms.

- *Supervised learning.* A classification model built using data instances of known origin. Once built, the model is able to determine the classification of instances of unknown origin.

- *Test set.* Data instances used to test supervised learning models.

- *Training data.* Data instances used to create supervised learning models.

- *Unsupervised clustering.* Machine learning models are built using an evaluation function that measures the goodness of placing instances into the same cluster. Data instances are grouped together based on a similarity scheme defined by the clustering model.

EXERCISES

Review Questions

1. Differentiate between the following terms:

   a. Training data and test data

   b. Input attribute and output attribute

   c. Supervised learning and unsupervised clustering

2. For each of the following problem scenarios, decide if a solution would best be addressed with supervised learning, unsupervised clustering, or database query. As appropriate, state any initial hypotheses you would like to test. If you decide that supervised learning or unsupervised clustering is the best answer, list two or more input attributes you believe to be relevant for solving the problem.

   a. What characteristics differentiate people who have had back surgery and have returned to work from those who have had back surgery and have not returned to their jobs?

   b. A major automotive manufacturer initiates a tire recall for one of their top-selling vehicles. The automotive company blames the tires for the unusually high accident rate seen with their top seller. The company producing the tires claims the high accident rate only occurs when their tires are on the vehicle in question. Who is to blame?

c. An adjustable bed company wants to develop a model to help determine proper settings for new customers purchasing one of their beds.

d. You desire to build a model for a fantasy football team to help determine which running back to play for a specific week.

e. When customers visit my Web site, what products are they most likely to buy together?

f. What percent of my employees miss one or more days of work per month.

g. What relationships can I find between an individual's height, weight, age, and favorite spectator sport?

3. List five or more attributes likely to be relevant in determining if an individual income tax filing is fraudulent.

4. Visit the Web site: https://en.wikipedia.org/wiki/Examples_of_data_mining and summarize five applications where machine learning has been used.

5. Go to the Web site www.kdnuggets.com.

a. Summarize one or two articles about how machine learning has been applied to solve real problems.

b. Follow the *Datasets* link and scroll the UCI KDD Database Repository for interesting datasets. Describe the attributes of two datasets you find interesting.

6. Search the Web to find a reference that gives the top ten machine learning algorithms. Provide a list of the names of these algorithms as well as the reference.

7. What happens when you try to build a decision tree for the data in Table 1.1 without employing the attributes *Swollen Glands* and *Fever*?

Computational Questions

1. Consider the following three-class confusion matrix. The matrix shows the classification results of a supervised model that uses previous voting records to determine the political party affiliation (Republican, Democrat, or Independent) of members of the United States Senate.

	Computed Decision		
	Republican	Democrat	Independent
Republican	42	2	1
Democrat	5	40	3
Independent	0	3	4

a. What percent of the instances were correctly classified?

b. According to the confusion matrix, how many Democrats are in the Senate? How many Republicans? How many Independents?

c. How many Republicans were classified as belonging to the Democratic Party?

d. How many Independents were classified as Republicans?

2. Suppose we have two classes each with 100 instances. The instances in one class contain information about individuals who currently have credit card insurance. The instances in the second class include information about individuals who have at least one credit card but are without credit card insurance. Use the following rule to answer the questions below:

IF *Life Insurance = Yes & Income > $50K*
THEN *Credit Card Insurance = Yes*
      Rule Precision = 80%
      Rule Coverage = 40%

a. How many individuals represented by the instances in the class of credit card insurance holders have life insurance and make more than $50,000 per year?

b. How many instances representing individuals who do not have credit card insurance have life insurance and make more than $50,000 per year?

3. Consider the confusion matrices shown below.

a. Compute the lift for Model X.

b. Compute the lift for Model Y.

Model X	Computed Accept	Computed Reject	Model Y	Computed Accept	Computed Reject
Accept	46	54	Accept	45	55
Reject	2245	7,655	Reject	1955	7,945

4. A certain mailing list consists of P names. Suppose a model has been built to determine a select group of individuals from the list who will receive a special flyer. As a second option, the flyer can be sent to all individuals on the list. Use the notation given in the confusion matrix below to show that the *lift* for choosing the model over sending the flyer to the entire population can be computed with the equation:

$$lift = \frac{C_{11}P}{(C_{11} + C_{12})(C_{11} + C_{21})}$$

Send Flyer?	Computed Send	Computed Don't Send
Send	C_{11}	C_{12}
Don't Send	C_{21}	C_{22}

## CHAPTER 2

Introduction to R

In This Chapter

- Navigating RStudio

- Basic R Functions

- Packages

- Helpful Information

CHAPTER 1 PRESENTED KEY concepts defining the fields of machine learning and data science. With these fundamentals in hand, it's time for you to become an active participant in an environment focused on learning by doing. The great news is that all of the tools needed for this exciting new venture are contained in the programming and statistical language known as *R*!

It is important to know that there is no prerequisite assumption of a background in computer science or computer programming. In Chapter 3, we do address the topic of programming in R, but expertise in writing computer programs is not needed to develop proficiency in machine learning with R. RStudio's friendly user interface provides all of the analytics capabilities you'll need without having to write large blocks of program code. Let's begin by answering the what and why of R!

2.1 INTRODUCING R AND RSTUDIO

R is an open-source language designed for general programming, statistical computing, graphics, and more. It is most often seen as a dialect of the *S* language developed at Bell Labs by John Chambers. Everything in R is considered to be an object characterized as either a function or a variable. To that extent, R can be loosely considered an object-oriented language. However, unlike most traditional object-oriented languages that focus on classes containing data fields and methods for manipulating the data in these fields, R's primary focus is the function. In fact, generic functions are one of the more interesting features of R as generic functions have the ability to behave differently depending on the type of data they receive.

One aspect of R that differentiates it from most traditional programming languages is its implementation as an interpreter. This allows us to directly interact with R as we do a calculator but with all of the features of a programming language. This fact as well as its functional nature, variable declarations without type, and support of recursive programming make it somewhat similar to the LISP programming language. Actually, a specific dialect of LISP known as Scheme was one of the inspirations in the development of R. As you work with R, you will soon discover the many things you can easily do with R that you never thought possible with any programming language!

2.1.1 Features of R

Here are a few of the many reasons why R is a good choice for your machine learning and data science projects.

- Unlike most sophisticated analytics tools that cost hundreds or thousands of dollars, R is free but has most if not more of the features of these expensive tools.

- We can interact with R using the RStudio Desktop IDE (integrated development environment) available in both desktop and server formats.

- In addition to a large set of standard packages, a growing database of new packages are available for you to freely download and use.

- R can be implemented on the three major platforms: Windows, Macintosh, and Unix.

- R supports sophisticated graphics capabilities.

- R integrates other software packages of which you may be familiar. For example, if you have worked with Weka, you will immediately feel right at home once you install the RWeka package.

- R can be integrated into other programming languages and analytics platforms of which you may already be familiar.

- A working knowledge of R is a common job requirement for data science professionals.

- R has a very large support community.

This is but a short list of features that make R an excellent tool for the beginning and seasoned student of machine learning and knowledge discovery. Let's put these features to work by installing R and RStudio!

2.1.2 Installing R

The latest version of R can be downloaded for free by clicking the download link at the Web site www.r-project.org shown in Figure 2.1. In the upper-left window under CRAN (Comprehensive R Archive Network) click on *download* which takes you to the CRAN

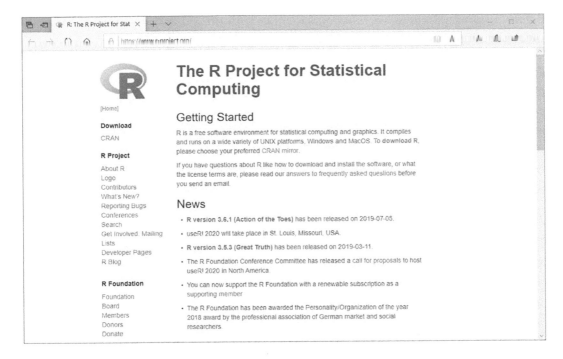

FIGURE 2.1   Obtaining and installing R.

download site. Scroll to find an appropriate site to download and install the latest version of R. Once installed, click the R icon on your screen and you will see R's graphical user interface (GUI) as in Figure 2.2. We could certainly use the R GUI for our work but, unless speed is of the utmost importance, the RStudio IDE is a much better choice.

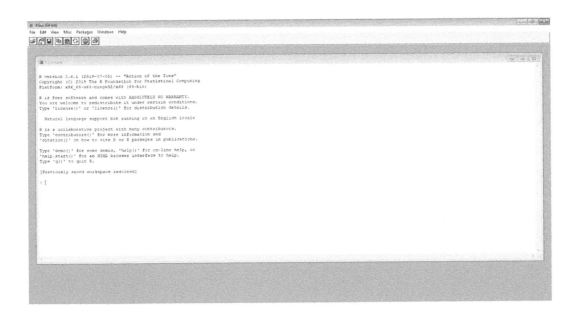

FIGURE 2.2   The R GUI.

Before installing RStudio, we must quit R either with the q function—type q( ) after the caret—or simply close the R GUI window.

2.1.3 Installing RStudio

RStudio is a commercial company that develops tools for R users. RStudio offers a free version—both desktop and server implementations—of their RStudio IDE product. To obtain the free desktop version, visit https://rstudio.com/ and click on *Download.* Your screen will appear similar to Figure 2.3. Click the free download link found directly under *RStudio Desktop.* When installation is complete, an RStudio icon will be placed on your screen. Click on the icon to have your screen appear as in Figure 2.4, minus the arrows!

2.2 NAVIGATING RSTUDIO

The RStudio IDE is divided into the four main regions shown in Figure 2.4. Let's take a look at the function of each region.

2.2.1 The Console

Region 1 is known as the console. Starting an R session with RStudio places you in the console where you see a message indicating the version of R, how to start a demo, access help, and how to quit. To clear the console, hold down the ctrl key and press L (ctrl-L). The blinking caret indicates the console is ready for input. Additional information about useful keystrokes and selected functions is given in Table 2.1. Of primary interest is the *Esc* key which, when all else fails, allows you to start over!

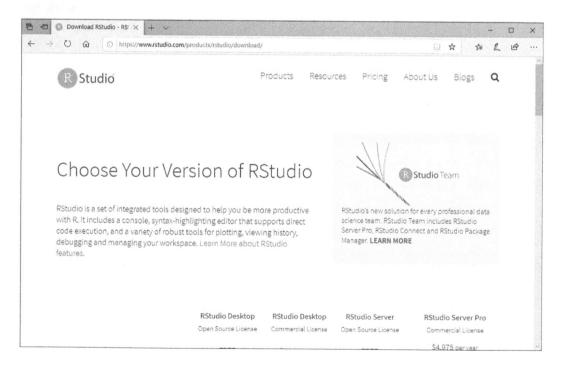

FIGURE 2.3   Obtaining and installing RStudio.

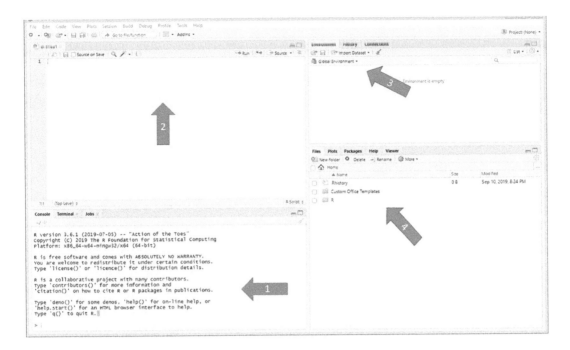

FIGURE 2.4   The RStudio interface.

TABLE 2.1   Common Keystrokes and Actions

Ctrl-L	Ctrl-L clears the console and places the caret at the top of the screen.
↑	Use up arrow to scroll through previous statements.
Esc	Press to terminate the currently running process.
#	Start each comment line with this symbol. There is no mutliline comment symbol.
Ctrl-shift-enter	Run an entire source file script.
Ctrl-shift-n	Begin creating a new script.
var <- 3.14	The left arrow followed by a minus sign is the most often used method for basic assignment. Here *var* is assigned the value 3.14.
quit( ), q ( ), or ctrl-q	Three ways to quit an R session.
rm(x)	Remove object x from the global environment.
ls( )	Lists all objects in the current workspace.
list.files()	Lists all files in your working directory.

Let's begin by typing the arithmetic expression 57 / 3 + 5 into the console window to see what happens

```
> 57 / 3 + 5
[1] 24
>
```

The result is displayed after a [1] and a new caret ready for input appears.

Let's try something a bit more interesting. The assignment statement in R is a left arrow followed by a minus sign. The equal sign (=) works, but you will not often see it used. Here we assign 3.14 to data object x.

```
> x<- 3.14
> x
[1] 3.14
```

The combine function *c( )* is used for vector assignment. Here the previous value of x is overwritten and x becomes a vector object with values 1–30.

```
> x<- c(1:30)
> x
[1]  1  2  3  4  5  6  7  8  9 10 11 12 13 14 15 16 17 18 19 20 21 22 23 24
25 26 27 28 29
[30] 30
```

We can easily reference any item within vector x. Here we reference the fifth item.

```
> x[5]
[1]  5
```

We often see datasets with at least some missing values. In R, missing items—both numeric and categorical—are represented by the symbol NA (not available). Here's an example:

```
> x[31]
[1]  NA
```

Methods for dealing with missing data are addressed in Chapter 4.

Let's assign my.sd the standard deviation of the values in x using the *sd* function. Unlike languages such as C++ and Java, a period in a statement has no special significance.

```
> my.sd <- sd(x)
> my.sd
[1] 8.803408
```

Here is a reassignment of my.sd using the *round* function.

```
> my.sd <- round(my.sd,2)
> my.sd
[1] 8.8
```

We can certainly continue in this fashion, but we are likely to get nowhere fast. We need a better method for taking care of business. A more efficient way of sequencing our actions is found within the source panel.

2.2.2 The Source Panel

Region 2 represents the source panel. This is where most of the action takes place. Here we load, view, edit, and save files; build scripts; and more! All scripts, datasets, and functions written especially for your text are part of the supplementary materials available to you.

If you have yet to acquire the supplements, turn to Appendix A as it shows how to download these materials.

2.2.2.1 The Power of Scripting

All R scripts are designated with a file name followed by a .R file extension. In addition, when referencing a path, you must use a forward rather than the typical backward slash.

Let's load a script designed to give insight into how scripting makes life much easier for us. To begin, locate Script 2.1 within your supplements and open it in your source editor.

The statements and output of Script 2.1 are shown in the listing below titled *Script 2.1 Statements in R*. To execute the script one line at a time, locate and click on *run*—upper right corner of the source editor. In doing so, you first see the output of *getwd( )*. This function gives the location of your working directory. The working directory is the default search path for locating files and is where your workspace history is stored. *setwd* changes the working directory to the location you specify. For simplicity, we have changed the working directory to the desktop.

When you exit RStudio, you will be asked if you want to save the *workspace image*. If you click on *save*, a history file—.RHistory—that can be accessed with any standard editor is created in you working directory. Also, all data objects and user functions you have introduced during your RStudio session will be saved. Unless you decide to remove these objects, they will remain from session to session as part of your global working environment. Please note that at the start of each RStudio session, the working directory resorts back to the original default directory. To change the default directory to better meet your needs, simply click on *tools* and then on *global options* to set a new default value.

As you continue to execute the individual script lines, notice that in addition to seeing output in your console, data objects are added one by one to the global environment—Region 3.

Also, *cat* and *print* are two alternatives for displaying output, *pi* is a reserved object, and the *mode* function returns data type—logical, numeric, complex, or character. Notice the mode of *y* changes from *logical* to *numeric* once we decide to add 1 to *y*! The *help* function displays helpful information about the *cat* function in the area designated as region 4.

Script 2.1 Statements in R

```
> getwd()
[1] "C:/Users/richa/Documents"
> setwd("C:/Users/richa/desktop")
> getwd()
[1] "C:/Users/richa/desktop"

> x <- 5+10        # assign a value to x
> print(x)         # Print to the console
[1] 15
> radius <- 3
> area<-pi*radius^2
> area
[1] 28.27433
```

```
> area <- round(area,2)
> area
[1] 28.27
> cat("area=", area)
area= 28.27
> y <-TRUE
> mode(y)
[1] logical
> y <- TRUE + 5
> mode(y)
[1] "numeric"
> cat(y)              # Print to the console
6
> x<- c(1:10)
> x
 [1]  1  2  3  4  5  6  7  8  9 10
> mean(x)
[1] 5.5
> median(x)
[1] 5.5
> length(x)
[1] 10
> # For help type help with the name of the function
> help(cat)
```

Finally, when you are ready to try your hand at building a script, mouse to *file, new file, R script*, or just type ctrl-shift-n. To save a script you have created, be sure to add the .R extension to the file name. RStudio considers a file without this extension a text file and some of the features given for script files will not be available.

2.2.3 The Global Environment

Region 3 holds references to the data and functions we have brought into the global environment.

A click on *History* provides a view of our recent workspace history. More importantly, although several read functions for importing data are available, a click on *Import Dataset* is the easiest way to bring a dataset into the global environment.

To see how functions work within the global environment, let's import the ccpromo. csv dataset found in your supplementary materials. The dataset is described in the box titled *The Credit Card Promotion Database*. The 15 instances within the data are shown in Table 2.2. Here's how to use the *import data* option.

2.2.3.1 Importing Data into the Global Environment

Click on *Import Dataset* and select the first option—from text (base….)—which tells R you are importing a text file. Figure 2.5 shows the import screen resulting from the request. Be sure *yes* is checked for *Heading*. This informs R that the first row of data signifies column names.

THE CREDIT CARD PROMOTION DATABASE

Even with the advent of paying bills on the Internet, more than 50% of all individual credit card holders still receive and pay their monthly statement through the postal service. For this reason, credit card companies often include promotional offerings with their monthly credit card billings. The offers provide the credit card customer with an opportunity to purchase items such as luggage, magazines, and jewelry. Credit card companies sponsoring new promotions frequently send bills to those individuals that have no current card balance hoping that some of these individuals will take advantage of one or more of the promotional offerings. From a machine learning perspective, with the right data, we may be able to find relationships that provide insight about the characteristics of those individuals likely to take advantage of future promotions. In doing so, we can divide the pool of zero-balance card holders into two classes, where one class will be those persons likely to take advantage of a new credit card promotion. These individuals should be sent a zero-balance billing containing the promotional information. The second class will consist of persons not likely to make a promotional purchase. These individuals should not be sent a zero-balance monthly statement. The end result is a savings in the form of decreased postage, paper, and processing costs for the credit card company.

The credit card promotion database shown in Table 2.2 has fictitious data about 15 individuals holding credit cards with the Acme Credit Card Company. The data contains information obtained about customers through their initial credit card application as well as data about whether these individuals have accepted various past promotional offerings sponsored by the credit card company. The dataset is in .csv format and titled *ccpromo*. Although the dataset is small, it serves well for purposes of illustration. We employ this dataset for descriptive purposes throughout the text.

TABLE 2.2   The Credit Card Promotion Database

ID	Income Range ($)	Magazine Promotion	Watch Promotion	Life Insurance Promotion	Credit Card Insurance	Gender	Age
1	40–50 K	Yes	No	No	No	Male	45
2	30–40 K	Yes	Yes	Yes	No	Female	40
3	40–50 K	No	No	No	No	Male	42
4	30–40 K	Yes	Yes	Yes	Yes	Male	43
5	50–60 K	Yes	No	Yes	No	Female	38
6	20–30 K	No	No	No	No	Female	55
7	30–40 K	Yes	No	Yes	Yes	Male	35
8	20–30 K	No	Yes	No	No	Male	27
9	30–40 K	Yes	No	No	No	Male	43
10	30–40 K	Yes	Yes	Yes	No	Female	41
11	40–50 K	No	Yes	Yes	No	Female	43
12	20–30 K	No	Yes	Yes	No	Male	29
13	50–60 K	Yes	Yes	Yes	No	Female	39
14	40–50 K	No	Yes	No	No	Male	55
15	20–30 K	No	No	Yes	Yes	Female	19

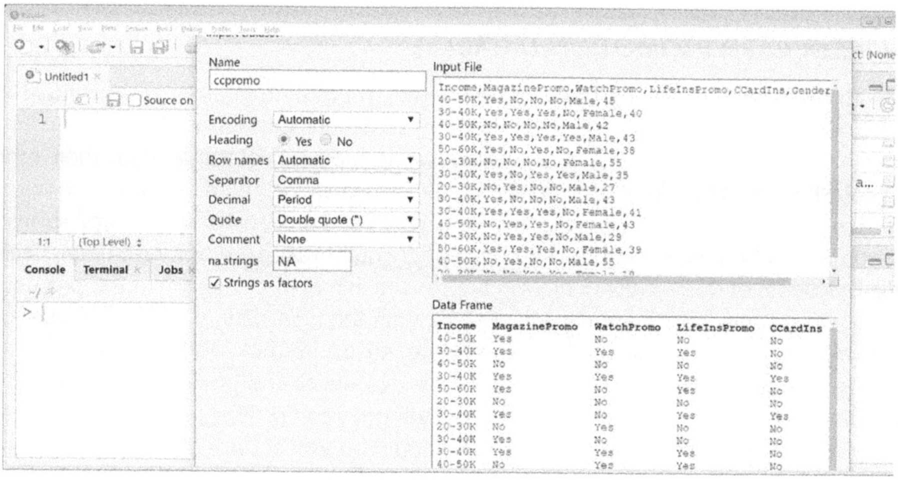

FIGURE 2.5   Importing ccpromo.csv.

Also, be sure *Strings as factors* is checked. The reason for this will be apparent a bit later. Click *import* to see a screen appear similar to Figure 2.6. Your global environment should now contain an object labeled *ccpromo*. Within this environment, *ccpromo* is stored as a structure known as a data frame. A data frame is a general two-dimensional structure that allows columns to contain different data types. You will learn more about data frames in Chapter 3.

Before working with *ccpromo*, it is instructive to know that most of the files in your supplementary dataset zip file are stored in both .csv and Microsoft Excel formats. As there can be issues when reading Excel files, the .csv format is highly recommended for file imports. The reason for maintaining the Excel versions of the files is that they contain documentation explaining the individual files and provide an easy way to recreate the corresponding .csv files. Let's take a closer look at the *ccpromo dataset* as it appears on your screen and in Figure 2.6.

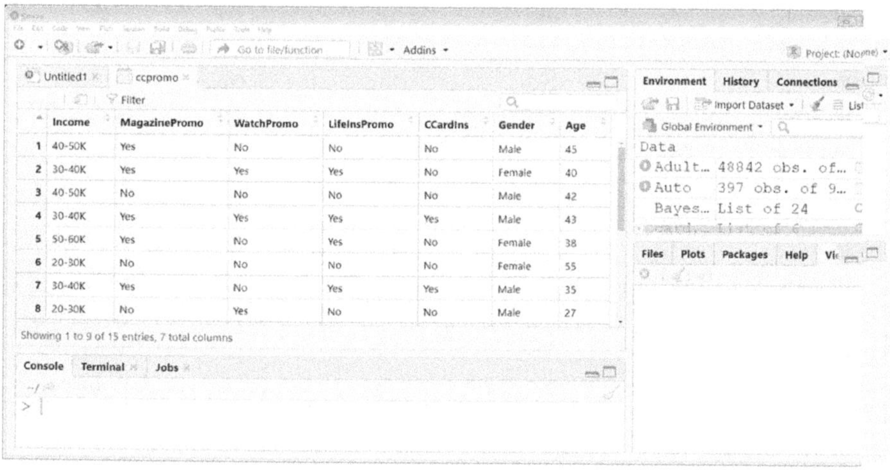

FIGURE 2.6   An imported data file.

2.2.3.2 Analyzing Credit Card Promotions

We created Script 2.2 to help explain several useful R functions for manipulating data. Load Script 2.2 into your source editor to better follow our explanation of the statements and output shown in *Script 2.2 Useful Functions*.

Script 2.2 Useful Functions

```
> ccp <- ccpromo
> summary(ccp)   # summary of the data
   Income  MagazinePromo WatchPromo LifeInsPromo CCardIns  Gender
 20-30K:4 No :7         No :7      No :6        No :12    Female:7
 30-40K:5 Yes:8         Yes:8      Yes:9        Yes: 3    Male  :8
 40-50K:4
 50-60K:2


        Age
 Min.   :19.0
 1st Qu.:36.5
 Median :41.0
 Mean   :39.6
 3rd Qu.:43.0
 Max.   :55.0


> nrow(ccp)      # Number of rows
[1] 15
> ncol(ccp)      # Number of columns
[1] 7


> head(ccp,3)    # Print first 3 rows
   Income MagazinePromo WatchPromo LifeInsPromo CCardIns Gender Age
1 40-50K           Yes         No          No        No   Male  45
2 30-40K           Yes        Yes         Yes        No Female  40
3 40-50K            No         No          No        No   Male  42
> table(ccp$Gender)
   Female    Male
        7       8


> table(ccp$Gender, ccp$LifeInsPromo)

         No Yes
  Female  1   6
  Male    5   3


> table(ccp$Income,ccp$LifeInsPromo,ccp$Gender)
, ,  = Female                    , ,  = Male
```

```
            No  Yes                     No Yes
   20-30K    1    1        20-30K    1    1
   30-40K    0    2        30-40K    1    2
   40-50K    0    1        40-50K    3    0
   50-60K    0    2        50-60K    0    0
> hist(ccp$Age,5)              # Use hist to create a histogram
> plot(ccp$Age,type='h')   # Use plot to create a histogram

> # Using the with statement
> with(ccp, {
+ print(table(Gender, LifeInsPromo))
+ print(table(Income,LifeInsPromo,Gender))
+ hist(Age,5)              # Use hist to create a histogram
+ plot(Age,type='h')    # Use plot to create a histogram
+ })
```

The first statement in Script 2.2 makes a copy of *ccpromo* to use throughout the script. This prevents having to reload *ccpromo* if the original copy becomes altered. *summary* is a generic function as it performs differently depending on its argument. In this case, it gives a summary of the attributes of *ccp*. Next, we obtain the number of rows (*nrow*) and the number of columns (ncol) in *ccp*. The function *head* lists the first three instances within the data—the default is six. By default, *tail* returns the last six items of *ccp*. The *table* function is first used here to obtain summary information about *gender*. Its second application creates a contingency table showing the relationship between *gender* and whether a life insurance promotional offer was accepted or rejected. The $ is likened to the decimal point in several programming languages as it specifies attribute names of interest within *ccp*. Using an attribute name without some form of data frame reference results in an error.

For the third example with the *table* function, a three-way table summarizes the relationship between *gender, income*, and acceptance or rejection of a life insurance promotional offering.

The next two statements in Script 2.2 create histograms for *age*. The first uses the *hist* function—Figure 2.7—and specifies five divisions for age. The second applies the *plot*

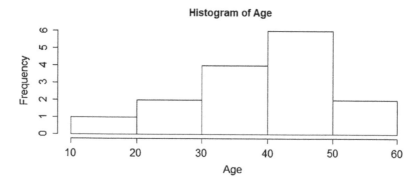

FIGURE 2.7 Histogram for age.

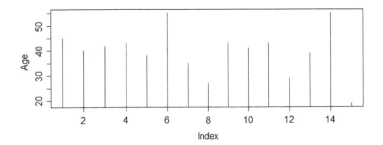

FIGURE 2.8   Histogram for age created with the plot function.

function—Figure 2.8—where index represents instance number. Be sure to use *Zoom* for a better depiction of your plots.

The last lines of Script 2.2 show how the *with* function avoids prefixing attribute names with their parent data frame. All attributes referenced in the *with* structure are automatically prefixed by *ccp*. A second possibility for avoiding a prefix makes use of the *attach* function. Invoking *attach(ccp)* adds *ccp* to R's search path, thereby making the ccp reference unnecessary. You can see the search path by typing *search ( )*.

A problem often seen when using *attach* is known as masking. Consider the example below where we use *attach* with the Cardiology dataset described in Chapter 1. Recall this data has an attribute named *class*.

```
> attach(CardiologyMixed)
The following object is masked _by_ .GlobalEnv:

    class
```

The message tells us there is already an object named *class* in the search path. This means we must reference the *class* attribute found in CardiologyMixed using the parent data frame. This can become confusing and quite problematic. The *with* function avoids this issue but its syntax can be a burden. If you choose to use *attach*, it is best to use *detach(name)* when you are done as it removes *name* from the search path.

2.2.4 Packages

A return to Figure 2.4 takes us to region 4 where we can search help, view graphs but most importantly install and update packages. Each package groups together a set of functions, datasets, and documentation for our use within the R environment. Approximately 30 packages are part of the base R distribution. In addition, several thousand packages are available for installation through the CRAN library. Packages are installed with the *install.packages( )* function which can be accessed via point and click within region 4. Installed packages are stored on your computer in a directory called *library*. The *library( ) function* returns a list of your installed packages.

Let's install two packages from the CRAN library using the point and click option. You will use functions from these packages in the chapters that follow. Let's first install *RWeka*.

In region 4, click on packages then on *install*. Type *RWeka* in the search bar. Once the package has been located click *install*. Upon completion, *RWeka* will appear in the list of installed packages. To install the second package click on *install*. Type *RWeka.jars* in the search bar to locate the package within the CRAN library. Once again, click *install*. Next, scroll the list of packages until you find both of the newly installed packages. Notice that although the packages are available, they have yet to be loaded. To load a package, place a check in the box by the package name or use the *library* function. For example, *library(RWeka)* loads the *RWeka* package thereby making the contents of the package available. To see what the *RWeka* package has to offer, click on the package name. Scrolling your screen gives you access to the documentation and help pages for the package.

2.3 WHERE'S THE DATA?

Most of the datasets used for the examples in your book are found in the supplementary materials. However, there is an abundance of data available within the R environment, as well as on the Internet. Appendix A gives information about obtaining the supplementary materials for your text and offers several Web sites for attaining additional data.

You can obtain a list of the datasets available with your currently loaded R packages by typing *data( )*. Let's add a dataset to the list. As the *AdultUCI* dataset is one of the more referenced datasets for machine learning, it is a viable choice. The dataset is made up of 15 attributes and close to 49,000 instances representing census information extracted from a government database. The dataset is located in the *arules* package which we will be using in Chapter 7. To access the data, we must first install the *arules* package. Here is the sequence of steps for installing the *arules* package from the console window and employing the table function to extract information from the AdultUCI data.

```
> install.packages("arules")# Install arules from the CRAN
Website.
> library(arules)        # Load the arules package.
> data(AdultUCI)         # Make the dataset available to us.
> View(AdultUCI)         # Examine the data in the source editor.
> attach(AdultUCI)       # Adds AdultUCI to the search path.
> table(sex,race)        # Create a simple table.
```

	race Amer-Indian-Eskimo	Asian-Pac-Islander	Black	Other	White
sex					
Female	185	517	2308	155	13027
Male	285	1002	2377	251	28735

The end-of-chapter exercises provide further opportunity for you to explore this data.

2.4 OBTAINING HELP AND ADDITIONAL INFORMATION

You may have noticed that after typing the first three characters of a function name in the source or console windows, a small window showing the function call together with available options appears. The window asks you to type F1 for additional help. If you oblige, the package interface (region 4) area gives the function name—with corresponding package name listed in parentheses—together with several forms of documentation. You can obtain this same information with *help(function-name)* or more simply *?function-name*.

Several online resources provide additional support. Two of note include:

- https://www.rdocumentation.org/—This site was developed by DataCamp and provides a wealth of information about the latest packages. If you search the Web for help on a particular topic, you will often find a response associated with DataCamp.

- www.r-consortium.org—The RConsortium consists of data scientists whose primary goal is to support the growth of the R community.

- https://support.rstudio.com/hc/en-us—This site offers support and information about RStudio.

2.5 SUMMARY

This chapter introduced the *R* programming language, a powerful analytics tool containing algorithms for data preprocessing, supervised learning, clustering, association rule mining, visualization, statistical analysis, and more. We also demonstrated how RStudio provides a friendly user interface for working with R. In the next chapter, we examine in detail the data types and data structures defining the R language. In addition, Chapter 3 offers the basics about how to write you own R functions!

EXERCISES

Please note: In addition to your output appearing in the console, you can use the *sink* function to write your output to an external file. Here's an example of how to use *sink*.

```
sink("C:/Users/richa/desktop/my.output", split = TRUE)

sink( )                 # returns output to console
```

1. Import the patient cardiology dataset described in Chapter 1 found in your supplementary materials as CardiologyMixed.csv. Perform the tasks below using this data. Recall that the data contains 303 instances representing patients who have a heart condition (sick) as well as those who do not.

   a. How many of the instances are classified as healthy?

   b. What percent of the data is *female*?

   c. Are there any instances with missing data? If so, what are the attributes containing missing items?

   d. What is the most commonly occurring value for the attribute *slope*?

   e. State two ways to obtain mean and median values for a numeric attribute. What is the mean and median *age* within the dataset?

f. What is the range (highest and lowest) and standard deviation for *maximum heart rate*?

g. How many instances have the value 2 for the *number of colored vessels*?

h. Display two histograms for *maximum heart rate*. Use the default number of intervals for one histogram and use five intervals for the second.

i. Use the *head* function to display the first ten values for *chest.pain.type*.

j. How many healthy individuals have a value of normal for *thal*? How many from the sick class have a value of normal for *thal*?

2. Use the AdultUCI dataset to answer the following.

a. What is the mean and median age within the data?

b. Are there any instances with missing data (NAs)? If so, what attributes have missing items?

c. What occupation is listed most frequently?

d. What is the mean and median hours worked per week?

e. What percent of the instances have a listed income of "small"?

f. What percent of the instances are listed as unmarried white females?

g. Create histograms for *hours per week* and *education num*. What do you learn from each graph?

3. Use the *help* function to specify how the functions *print* and *cat* differ. Give an example of using each with the CardiologyMixed data object.

Installed Packages and Functions

Package Name	Function(s)
arules	**
base / stats	attach, c, cat, data, detach, getwd, head help, hist, install. packages, length, library, list.files, ls, mean, median, mode, ncol, nrow, plot, print, q, quit, rm, round, setwd, sink, summary, table, tail, View, with
RWeka	**
RWeka.jars	**

## CHAPTER 3

Data Structures and Manipulation

In This Chapter

- Data Types
- Data Structures
- Data Manipulation
- Writing Your Own Functions

IN THIS CHAPTER, we broaden our investigation of R. We start in Section 3.1 by detailing the data types supported by R. In Section 3.2, we examine the single-mode data structures available with R. In Section 3.3, our attention is on multimode data structures with special emphasis placed on the role of data frames. Several data manipulation techniques are given here with more introduced throughout the text. Section 3.4 presents a quick guide to programming in R. Your interest may not lie in learning to write your own R functions. However, many R scripts contain at least a line or two of R program code. A short review of this section provides a basic understanding of the most frequently used R programming structures.

3.1 DATA TYPES

R supports four main types of data: numeric, character, logical, and complex. The *mode* function tells us the data type for any value. Here are examples of each data type:

```
> mode(4)
[1] "numeric"
> mode(4.1)
[1] "numeric"
```

```
> mode('cat')
[1] "character"
> mode("dog")
[1] "character"
> mode(TRUE)
[1] "logical"
> mode(3 + 4i)
[1] "complex"
```

As you can see, numbers both with and without a decimal point default to a numeric representation. Also, character values can be identified with either single or double quotes. Although integer is not a basic data type in R, we can represent numbers as integers. Consider these statements:

```
> is.integer(4)
[1] FALSE
> y <- as.integer(4)
> is.integer(y)
[1] TRUE
> y==4.0
[1] TRUE
```

By default, 4 is stored as a decimal (floating point) value. So the is.integer test fails. However, we can store 4 as an integer with the as.integer function. When we test to see if integer y and decimal 4 are the same, we see TRUE as R first converts y to decimal in order to make the comparison. TRUE and FALSE are of type logical; however, they are stored internally as 1 and 0, respectively. This allows their use for arithmetic computations.

```
> TRUE + TRUE
[1] 2
> FALSE + TRUE
[1] 1
> FALSE - 5
[1] -5
```

Although an interesting topic, data type doesn't often cause us problems. However, there can be an issue when all of the data for a particular attribute is integer, but there is the possibility of seeing a decimal value for the attribute at a later time. We will look at how to handle this situation in Chapter 6.

3.1.1 Character Data and Factors

Character data can be represented as numeric vectors known as *factors*. The advantage is twofold. First, numeric storage for strings of characters is memory efficient. For example, suppose we have a dataset of 50,000 instances where one column of character data represents health status with each instance showing either "healthy" or "sick". If we store the column of data as a factor variable, we can assign a 1 to healthy and a 2 to sick. We must also have a chart that tells us how the initial values were assigned. With this scheme, we are able to replace all values of "healthy" and "sick" with 1's and 2's. At the same time, if we query any instance in the dataset, we have the chart to tell us the health status of the individual in question. Let's see how this works in R.

Consider the following statements:

```
> my.result <- c('good','bad','not sure',
+                'bad','good','good','not sure')
> str(my.result)
 chr [1:7] "good" "bad" "not sure" "bad" "good" "good" "not sure"
> my.result <- factor(my.result)
> my.result
[1] good      bad      not sure bad      good     good     not sure
Levels: bad good not sure
> str(my.result)
 Factor w/ 3 levels "bad","good","not sure": 2 1 3 1 2 2 3
```

The first statement assigns a vector of character strings to *my.result*. Next, we see a call to the *str* function. This function when given a data object displays its internal format. Here it tells us *my.result* is a vector of seven items having type *chr*.

Next, the *factor* function converts the vector of character strings to a factor of three levels. The three levels represent the original character strings. The *str* function also shows the numeric values that substitute for the character strings. Here's where we must be careful!

By default, the numeric assignment of numbers to character strings is done alphabetically. That is, 1 is assigned to "bad", 2 to "good", and 3 signifies "not sure". After the displayed factor levels, we have the string of numbers, 2 1 3 1 2 2 3. We first see 2 which matches with "good". "Good" is also the first item listed in the original vector. The original vector lists "bad" in the second position, so we have a 1. The 3 tells us the third position in the original list is "not sure", and so on it goes.

If you don't like the default assignment of numbers to factors, it can be changed as follows:

```
> my.result <- c('good','bad','not sure',
+                'bad','good','good','not sure')

> my.result<- factor(my.result, order =TRUE,
+                levels=c('good','not sure','bad'))
> str(my.result)
 Ord.factor w/ 3 levels "good"<"not sure"<..: 1 3 2 3 1 1 2
```

We now have "good" assigned to 1, "not sure" is a 2, and 3 is "bad".

The *str* function is a very handy tool to apply in many situations, especially when we see error messages stating the input data is not formatted correctly. Let's apply the *str* function to the *ccpromo* dataset from Chapter 2. Your global environment should still have *ccpromo* listed as a data object. Here is the result.

```
> str(ccpromo)          # Internal storage structure
'data.frame': 15 obs. of  7 variables:
 $ Income        : Factor w/ 4 levels "20-30K","30-40K",.: 3 2 3 2 4
 $ MagazinePromo: Factor w/ 2 levels "No","Yes": 2 2 1 2 2 1 2 1 2 2
 $ WatchPromo    : Factor w/ 2 levels "No","Yes": 1 2 1 2 1 1 1 2 1 2
 $ LifeInsPromo  : Factor w/ 2 levels "No","Yes": 1 2 1 2 2 1 2 1 1 2
 $ CCardIns      : Factor w/ 2 levels "No","Yes": 1 1 1 2 1 1 2 1 1 1
 $ Gender        : Facto r w/ 2 levels "Female","Male": 2 1 2 2 1 1 2
 $ Age           : int   45 40 42 43 38 55 35 27 43 41 .
```

The "data.frame" declaration tells us the file is a data frame having 15 instances and 7 attributes. Most of R's machine learning tools require data to be structured as a data frame (see Section 3.3). We also have six of the seven attributes within the data declared as factors. All of the *No, Yes* attributes show *No* assigned to 1 and *Yes* to 2. This corroborates factor levels defaulting to an alphabetical assignment. Finally, since all age values are integer, *age* is given an integer data type.

3.2 SINGLE-MODE DATA STRUCTURES

R supports three types of single-mode data structures: vectors, arrays, and matrices. These are known as single-mode structures as each data entry must be of the same type. Let's take a look at each structure.

3.2.1 Vectors

In Chapter 2, we demonstrated how the combine function *c( )* is used to create vectors. Vectors are single-dimension structures. R performs automatic type conversions to keep the single type rule from being violated. To see this, consider the following where numeric vector *x* becomes a character vector then a factor.

```
> x<- c(1:5)        #create a vector of 5 numeric values
> x
[1] 1 2 3 4 5
> x[7]<- "run"     # Add a character item at position 7
> x
[1] "1"    "2"    "3"    "4"    "5"    NA      "run"
> str(x)
 chr [1:7] "1" "2" "3" "4" "5" NA "run"
> x
> x<-factor(x)      # Change characters to factors
> x
[1] 1      2      3      4      5      <NA> run
Levels: 1 2 3 4 5 run
```

3.2.1.1 Vectorization

An interesting feature of R is function *vectorization*. A function with this capability automatically performs its task over each item within a vector. Here are some examples:

```
> x<- c(10,-20,30,-40,50,-60)
> abs(x)
[1]  10 20 30 40 50 60

> sqrt(x)
[1] 3.162278      NaN 5.477226      NaN 7.071068      NaN
Warning message:
In sqrt(x) : NaNs produced

x>20
[1] FALSE FALSE TRUE FALSE TRUE FALSE

y<- x[x>20]
> y
[1]  30 50


>Over35 <- ifelse(AdultUCI$`hours-per-week` >35,TRUE,FALSE)

> table(Over35)
Over35
FALSE   TRUE
10332 38510

colSums(ccpromo[7])   # Find the sum of all ages in the data.
Age
594
```

The vectorization of the *ifelse* structure is of particular interest. In the example above, we have applied *ifelse* to the adult dataset. The statement works like this: For each instance, look at the value of *hours-per-week*; if *hours-per-week* is greater than 35, output TRUE otherwise output FALSE. The data object *Over35* contains the results which are summarized with the *table* function.

If you have done some programming, you can certainly relate to the simplicity and power of the *ifelse* statement! You will often see *ifelse* used to convert a probability-based outcome to an integer or character value. In the example below, all values greater than 0.5 in *results* are represented in *new.results* as 2. Likewise, those values less than or equal to 0.5 are shown as 1 in *new results*.

```
new.results <- ifelse(results > 0.5,2,1)
```

These examples show that vectorization avoids having to write functions that loop through the data. Functions that have been vectorized are particularly helpful when dealing with large volumes of data where execution efficiency is a must.

3.2.2 Matrices and Arrays

A matrix is a two-dimensional structure with the requirement that all elements are of the same mode. We can create a matrix with the *matrix* function. By default, a matrix is filled in column first order. Two examples follow. The first example creates a four-row three-column matrix. The second example defines a 3×2 matrix with assigned names.

```
my.matrix <- matrix(1:12,nrow=4,ncol=3)
> my.matrix
      [,1] [,2] [,3]
[1,]    1    5    9
[2,]    2    6   10
[3,]    3    7   11
[4,]    4    8   12


> items <- c('bread','milk','cereal','grapes','bacon','fish')
> my.matrix2 <- matrix(items, nrow=3,ncol=2,
+            dimnames =list(c('r1','r2','r3'),
+            c('c1','c2')))
>  my.matrix2
     c1        c2
r1 "bread"   "grapes"
r2 "milk"    "bacon"
r3 "cereal" "fish"
> my.matrix2[1,]  # everything in row 1
     c1        c2
 "bread" "grapes"
> my.matrix2[,2]  # everything in column 2
     r1         r2         r3
"grapes"   "bacon"    "fish"
```

Arrays are similar to matrices in that all elements must be of a single mode. They differ from matrices in that they are not limited to two dimensions. Let's create a three-dimensional $2 \times 2 \times 2$ array. Here is the technique:

```
my.array <- array(1:8, c(2,2,2), dimnames=list(c("A1","A2"),
+                                          c("B1","B2"),
+                                          c("C1","C2")))
> my.array
```

```
, , C1

    B1  B2
A1   1   3
A2   2   4

, , C2

    B1  B2
A1   5   7
A2   6   8
```

Arrays beyond three dimensions are difficult to visualize. Like matrices, the default is to fill the arrays in column first order. Arrays are useful for storing large amounts of related information. However, from a machine learning perspective, the requirement that all elements of an array must be of the same data type makes them of limited use. It's time to proceed to multimodal data types where we find the data structures most often applicable for machine learning applications.

3.3 MULTIMODE DATA STRUCTURES

3.3.1 Lists

Lists are very general structures that are containers for other objects. Unlike vectors, matrices, and arrays, the objects in a list can be of varying type and mode. Lists are best understood by example. Here's how to create a list with information about my dog Lieue.

```
my.dog <- list(Name = "Lieue",
+                age = 14,
+                weight = 48,
+                breed= 'springer spaniel',
+                WeeklyWalk =c(2,2,2.5,1,0.5,3.1,0)) #end

> my.dog # Show the contents of the list.
$Name
[1] "Lieue"

$age
[1] 14

$weight
[1] 48

$breed
[1] "springer spaniel"
```

```
$WeeklyWalk
[1] 2.0 2.0 2.5 1.0 0.5 3.1 0.0

> mean(my.dog$WeeklyWalk)        # Determine the average daily walk
length
[1] 1.585714
```

We can easily make changes to the list. Here are some examples:

```
my.dog[3]               # Extract the weight element
$weight
[1] 48
> my.dog[[3]]           # Extract just the weight
[1] 48
> my.dog[[3]]<- 50      # Increase in weight
> my.dog[[5]]<- c(2.4)  #Start a new week of walking
> my.dog$WeeklyWalk
[1] 2.4
```

We have barely scratched the surface of what we can do with lists, but as students of machine learning, our primary interest lies with the data frame!

3.3.2 Data Frames

Data frames are similar to matrices as they are two-dimensional (row–column) structures. They differ in that each column of a data frame can be a different data type. This makes the data frame an ideal structure for machine learning with rows representing instances and columns signifying attributes or variables. When importing a dataset, it is always a good idea to use the *str* function to make sure the dataset has been imported as a data frame.

We can create a new data frame or transform an existing object into a data frame with the *data.frame* function. Script 3.1 provides an example of how to create and update a data frame. Functions *cbind* and *rbind* are of particular interest. As you can see, *cbind* adds new columns of data and *rbind* adds new rows. Finally, the *subset* function allows us to easily access subcomponents of a data frame. Script 3.2 shows additional techniques for extracting subsets of data from a data frame.

Script 3.1 Creating and Updating a Data Frame

```
> Age <- c(45,40,42,43,38)
> Gender<- c("Male","Female","Male","Male","Female")
> my.df <- data.frame(Gender,Age)
> my.df
  Gender Age
1   Male  45
```

```
2 Female   40
3   Male   42
4   Male   43
5 Female   38
> CCardIns<- c("No","No","No","Yes","No")
> my.df <- cbind(CCardIns,my.df)      # Add a new column
> my.df
  CCardIns Gender Age
1       No   Male  45
2       No Female  40
3       No   Male  42
4      Yes   Male  43
5       No Female  38
> add1 <- c("No","Female",55)
> my.df <-rbind(my.df,add1)      # Add a new row
> my.df
  CCardIns Gender Age
1       No   Male  45
2       No Female  40
3       No   Male  42
4      Yes   Male  43
5       No Female  38
6       No Female  55


subset(my.df,select=c(1,3))

  CCardIns Age
1       No  45
2       No  40
3       No  42
4      Yes  43
5       No  38
6       No  55


subset(my.df,Age <40)

  CCardIns Gender Age
5       No Female  38
```

Script 3.2 is not executed here, but you can load it into your source editor to see the effects each statement has on the data. Subscripting methods similar to those displayed in the script are very common. Getting subscripting to do exactly as you expect does take some practice. Of particular note is the statement that removes columns 2, 3, and 4 by placing a minus sign in front of a vector containing the column numbers to remove. Without the minus sign, our output will be limited to the specified columns.

Script 3.2 Row and Column Manipulation

```
ccp.data <- ccpromo
ccp.data          # List entire data frame
ccp.data[1]       # Everything in column 1
ccp.data[1,]      # Everything in row 1
ccp.data[,1]      # Everything in column 1
ccp.data[1:5,]    # Rows 1 through 5
ccp.data[,1:5]    # Columns 1 through 5
ccp.data[1:3,1:3]                      # Columns and rows 1 through 3
ccpNew.data <- ccp.data[-c(2,3,4)]  # Remove columns 2,3,4
ccpNew.data
```

Finally, upon installing and loading the *sqldf* package, you have access to the function *sqldf* which allows you to perform SQL queries on data frames! Here is an example using the ccpromo dataset.

```
library(sqldf)
> sqldf("select LifeInsPromo,age from ccpromo where Age<30 ")
  LifeInsPromo Age
1           No  27
2          Yes  29
3          Yes  19
```

3.4 WRITING YOUR OWN FUNCTIONS

It's a rare occasion that an R script of some complexity doesn't have a line or two of program code. Many of the scripts written in your text are no exception. To this extent, it's important for you to obtain a reading knowledge of the R programming language. However, with thousands of R packages readily available, you are likely to find a package with the exact function you need to perform any task. Given that, is there a case for learning how to write your own functions? Absolutely!

If program code runs through your veins, R is the perfect language for doing your own thing. Writing your own code also puts you in control and who doesn't like to be in the driver's seat? But what if you are new to programming? One of the positive features of R is that it is relatively easy to learn. Not having to concern yourself with strict data typing and being able to interact directly with R through an interpreter makes the beginning programmer's life a lot easier. Most importantly, if you work in any area of technology, knowing how to read and understand some R program code is a real plus. Let's write a few functions to get you familiar with programming in R. This will help you decide if programming in R is right for you!

3.4.1 Writing a Simple Function

Here is the general structure of an R function:

```
my.function <- function(arg1,arg2.....,argn)
  {
```

```
  statement 1
  statement 2
  ...
  statement n
  return(value)
}
```

The first line specifies the name of the new function and shows the arguments passed to the function. The body of the function is surrounded by curly brackets with the last statement specifying a return value. Let's start simple by writing the function my.square that—given a number—returns its square. For example, a call to my.square(4) returns 16.

To create the function, click on *file—new file—R Script.* Type the following code into your source editor:

```
my.square <- function(x)
{
    # Return the square of x
  return(x*x)
}
```

Next, click *run* which adds the function my.square to your global environment. Provided all goes well, your console window will appear as below:

```
> my.square <- function(x)
+ {
+     # Return the square of x
+   return(x*x)
+ }
>
```

If the last line shows as a + rather than a >, the interpreter is indicating it expects additional input. This represents a coding error such as a set of mismatched parenthesis.

Now that we have my.square defined, let's see if it works! Here are a few function calls to test my.square.

```
> my.square(5)
[1] 25
> my.square(-5)
[1] 25
> my.square("dog")
Error in x * x : non-numeric argument to binary operator
> y <- 10
> my.square(y)
[1] 100
> y
[1] 10
```

```
> y <- my.square(y)
> y
[1] 100
> my.lst <- c(5,10,20)
> sapply(my.lst,my.square)
[1]   25 100 400
```

The third call to my.square results in the interpreter giving us an error. As we should be responsible for catching this error, we will modify the function to catch the error shortly. The fifth call is of special interest as we are using the function to change the value of y. R uses *call by value* when passing variables to a function. With *call by value* the function is given the value of the variable being passed but does not know its location in memory. In this way, changes made to the argument inside the function have no permanent effect on the variable. If we want the function to change the value of a parameter, we can use the technique shown here. The final example shows how function *sapply* is used to apply our function to each member of my.lst.

3.4.2 Conditional Statements

The third call to my.square resulted in a system error that we can easily catch within our function. All we need is a basic knowledge of conditionals. Here is a modification to the original code that catches the non-numeric value error.

```
 my.square <- function(x)
{
  if(is.integer(x) | is.numeric(x))
    {# Return the square of x
     return(x*x)
    }
  else
    {print("Error! Can't square a non-numeric value")}
}
```

The code above contains a conditional in the form of an if-else. If the test within the parenthesis is true, the code following the *if* statement executes. If the statement is false, the code following the *else* is executed. The vectorized *ifelse* structure described earlier is a special case of the general if-else conditional.

An *if* conditional can also exist without the else. In this case, when the *if* statement is true, the statements enclosed within the curly brackets following the *if* are executed. If the statement is false, the statements within the curly brackets are skipped. In this situation, the *if* is said to have a null *else* clause. It is worth noting that the curly brackets are only needed when more than one statement follows an *if* statement. In the above example, the curly brackets are optional.

We see two new functions, *is.integer* and *is.numeric*, for checking the data type of x. The "|" is the logical operator for an *or* conditional test. Table 3.1 shows the most common

TABLE 3.1   Logical Operators

Operator	Function
>	Greater than
>=	Greater than or equal to
<	Less than
<=	Less than or equal to
==	equal to (two = signs)
!=	not equal to
!x	not x
x | y	x or y
x & y	x and y

logical operators. If either statement of the *if* test is true, the function returns the squared value. If not, the else component executes and prints the error message.

Here is a way to eliminate the *else* but still accomplish the same task.

```
my.square <- function(x)
{
  if(is.integer(x) | is.numeric(x))
    # Return the square of x
    return(x*x)
  print("Error! Can't square a non-numeric value")
}
```

The modification above takes advantage of the fact that if the conditional is true, the function returns a value before the print statement has a chance to execute. If the statement is false, the error message is printed. As if-else structures can be nested, they often take on a more complex form.

3.4.3 Iteration

Programs of any complexity almost always use some form of repetition. Repetition comes in two general formats—iteration and recursion. Although today's programming languages support both types of repetition, most languages emphasize iteration in the form of for, while, and do-while loops. However, some languages such as LISP, Scheme, and Prolog to name a few are built for recursive programming.

R supports both iteration and recursion but is considered iterative in nature. Let's first look at two examples of iteration with R. We will follow with a simple example illustrating recursion. Please note that all of the R functions written specifically for this text are available in your supplementary materials in the file *functions.zip*. You are encouraged to open each function in RStudio, then click *run* to have the function stored as a functional object within your global environment. In this way, a script that references one or several of these functions won't result in an error.

3.4.3.1 Finding the Square Root

For our first example, we use a *while* loop to implement the guess-divide-average method for finding the square root of a positive number. The algorithm is straightforward. Suppose we want to find the square root of 6. First, we must make an initial guess. An initial guess of 2 can always be made without any loss of generality. With our initial guess of 2, we divide our guess into 6 and obtain a quotient of 3. Next, we compute the absolute difference between our guess (2) and the quotient (3). If the absolute difference is less than a chosen value such as 0.001, the guess—or quotient as their difference is within the error bound—is taken as the square root. If the difference is too large, we compute a new guess by averaging the old guess (2) and the quotient (3). In this example, the difference between the quotient and the guess is (3 − 2 = 1). Therefore, we continue with a new guess. The new guess is (3 + 2)/2 = 2.5. Next, we divide 2.5 into 6 giving 2.4. Assuming the chosen absolute error is less than 0.01, the process repeats with 2.45 as the new guess.

The algorithm—shown in Script 3.3—is implemented with two functions, the first of which checks to determine if the argument is valid. If not, an error message is printed and the process terminates. A valid test results in the first function calling the second function with the number and the initial guess.

Script 3.3 Finding the Square Root with Guess-Divide-Average

```
my.sqrt <- function (x)
{
  # Find the square root of a number using the
  # guess-divide-average method.
  if(is.numeric(x) & x>=0)
  {
    Guess = 2
    my.sqrt2(x,guess)
  }
  else
    print("Invalid parameter!")
} # end my.sqrt

my.sqrt2 <- function(x,guess)
{

  while(abs(guess- x/guess) > 0.001)
    {
      guess <-(guess + x/guess) / 2
    }
  return(guess)
  } # end my.sqrt2
```

The *while* loop is key to understanding how the algorithm works. The *while* loop tests the conditional statement and if true executes the statement(s) within the while. Here, if the absolute value of (guess − x/guess) is greater than 0.001, the statements within the while will execute after which the test is repeated. As the while loop contains but one statement, the curly brackets are not needed. However, good programming practice encourages the brackets as they clearly define the boundaries of the while loop.

When the *while* test fails, the last computed value for guess is returned to my.sqrt which in turn returns the value to the calling statement. Lastly, it's important to note that any variables defined within an R function or listed as parameters are local to the function and are destroyed once the function terminates. Here local variable *guess* and formal parameter *x* disappear when the function terminates.

When the number of iterations is known prior to program execution, a *for* loop is the preferred iterative technique. The next example shows how a *for* loop is used to find the largest value in a list of numbers.

3.4.3.2 Finding the Largest Value

Our second example illustrating iteration defines a function that takes as input a vector of numeric data and returns the largest value. Script 3.4 defines the function.

Script 3.4 Finding the Largest Value

```
my.largest <-function (x)
  # Find the largest value in vector x. The assumption is
  # the first item in x is not NA.
{
  largest <- x[1] # initialize the largest value
  nas <- 0         # Initialize the number of NAs

  rows = length(x)-1

  for (i in 2:rows)
  {
    if(is.na(x[i]))
      {
      nas=nas +1
      }
    else
      {
      if(x[i]> largest)
        largest <- x[i]
      }
  } # End for

  # Print the number of NAs
  cat("No. of NAs ",nas,"\n")
  # Return the largest value
    return(largest)
} # End my.largest
```

my.largest starts by declaring the first value in the list as the largest item. Next, local variable *nas* is declared to keep track of the number of missing items within the vector. As the iteration will start at 2, rows is declared to be one less than the length of the list. The *for* statement begins with *i* equal to 2.

The first *if* conditional checks for a missing item. The second *if* determines whether the value of largest must change. The curly brackets starting after the *for* declaration and ending with # *end for* define the loop. Variable *i* increments each time by 1 until its value exceeds the value of *rows*. When the iteration is complete, the total number of NAs is printed and the largest value is returned.

3.4.3.3 Confusion Matrix Accuracy

Our final example with iteration is a function we use throughout the text to compute classification accuracy based on the values in a confusion matrix. The function sums the numbers in the main diagonal of the confusion matrix and then divides this sum by the total number of values in the matrix giving the accuracy value. The function uses nested *for* statements to iterate over the rows and columns of the matrix. Before you examine the statements, you can see how a nested *for* works by typing the following into your console or source editor. The execution (not displayed) shows the outer loop iterates from one to three while the inner loop iterates from 1 to 3 for each value of *i*.

```
> for(i in 1:3)
+    for(j in 1:3)
+       cat("i=",i," j=",j,"\n")
```

Script 3.5 gives the definition of *ConfusionP*. Be sure to open *ConfusionP.R* in your source editor and click *run* to enter *ConfusionP* as a function in your global environment.

Script 3.5 Computing Classification Accuracy

```
confusionP <- function (x)
   # This function prints classification accuracy
   # based on the values in confusion matrix x.

{correct=0
 wrong =0
y<- nrow(x)
z<- ncol(x)
for (i in 1:y)
   {
   for (j in 1:z)
     {
     if(i==j)
        correct = correct + x[i,j]
     else
        wrong = wrong + x[i,j]
     } # end for j
```

```
    } # end for i
pc <- (round(correct/(correct + wrong)*100,2))
cat(" Correct=", correct,"\n")
cat("Incorrect=", wrong,"\n")

cat("Accuracy =",pc,"%","\n") }
```

3.4.4 Recursive Programming

Recursive functions have two main characteristics. First, the function must call itself. This is known as the recursive call. Second, there must be a condition to terminate the recursion.

Recursion is implemented internally with what is known as a *stack*. A stack is a last in first out structure (LIFO) where the last item placed on the stack is the first item removed. A stack of dinner plates makes the idea clear. Suppose you are drying plates and after drying each plate you place it on the top of a growing stack of clean plates. If someone is in need of a dinner plate, the plate removed is the last one placed on the stack. Obvious problems—such as the entire stack of plates crashing to the ground—arise when significantly more plates are placed on instead of taken off the stack.

The issue with recursive programming is similar. Each recursive call requires a new plate filled with information about the current state to be placed in memory. The information must be stored as the information is used to compute a final result when the recursion unwinds (the plates come off the stack one by one). If there are too many recursive calls, memory reserved for the stack becomes full and the stack overflows!

The types of problems that can be easily solved with recursion are those that require a minimal number of recursive calls and a minimal amount of stored information to track the recursion. The guess-divide-average square root technique lends itself naturally to a recursive solution. This is the case as the technique is always able to compute the square root value with a minimal number of computations. More importantly, there is no need to store information on the stack as the final answer only needs to be passed back through the previous recursive calls. That is, there are no partial computations to be satisfied as the recursion unwinds.

Script 3.6 gives a recursive version of the guess-divide-average technique. The function my.SqrtR is the main function called by the user, but it does not perform the recursion. It simply checks to make sure the value whose square root is to be computed is valid. If valid, the recursive routine my.SqrtR2 is called with this value and the initial guess. Notice that repetition is accomplished simply by continuing to call my.Sqrt2 with guesses closer and closer to the actual square root. There are no *for* or *while* loops in sight! When the terminating condition is satisfied, the recursion passes the computed square root back through the stack of recursive calls and finally to the user. To be sure, recursive programming isn't for everyone. We very briefly introduced it here for those interested in taking on the challenge!

Script 3.6 A Recursive Implementation of Guess-Divide-Average

```
my.SqrtR <- function (x)
{ # Find the square root using guess-divide-average
```

```
  # as implemented with recursion.

  if(x<0)
    print("Error!")
  else
    my.SqrtR2(x,x/2)
} # End my.SqrtR

my.SqrtR2 <- function(x, guess)
{
if(abs(guess - x/guess)< 0.001)
  {
      return(guess) # The recursion unwinds
  }
  my.SqrtR2(x, (guess + x/guess)/2.0) #push on the stack
  }
```

3.5 SUMMARY

R supports several single-mode data structures including vectors, matrices, and arrays. These structures are dubbed as single mode in that they require all elements to be of the same data type. If the restriction is violated, R automatically performs an attribute-value conversion. Lists and data frames are multimode data structures as they lift the data type restriction. The data frame is the most common data structure used for the machine learning algorithms in R. The rows of a data frame represent instances and the columns serve as attributes (variables). The *str* function is particularly useful for gaining a better understanding of the internal representation of the data structures supported by R. R scripts often contain some program code. A study of how to write your own R functions is highly recommended as it offers a better understanding of the scripts written by others and provides an additional tool for increasing your problem-solving skills.

3.6 KEY TERMS

- *Call by Value.* The called function receives the value of passed parameters but does not know their address in memory. Any variable passed as a parameter will not have its value permanently changed by the function.

- *Factor.* A numeric vector that represents character data. The factor has levels that define the mapping of numeric values to their original character representation.

- *Stack.* A stack is a LIFO structure where the last item placed on the stack is the first item removed.

- *Vectorization.* A vectorized function applies itself to each value within a vector, thereby eliminating the need for looping.

EXERCISES

1. Consider the following declaration:

   ```
   x<- matrix(1:12,nrow = 3,ncol = 4)
   ```

   Replace position [1,3] with the character "A" and output the matrix. What is the result?

2. Make a copy—noted as CRD below—of CardiologyMixed. Use the copy to write and execute a script to perform the tasks below.

   a. List the ages of the first ten individuals in the data (this one's done!)
      Answer:

      ```
      CRD[1:10,1]
      [1]  60 49 64 63 53 58 58 58 63 67
      ```

   b. Use two methods for printing the first five instances.

   c. Print the age, gender, and chest.pain.type for instances 1, 20, and 51.

   d. Use two methods for printing all data in columns 1, 3, and 5.

   e. Remove columns 1 through 10.

3. Open mySqrt.R in RStudio and make my.sqrt and my.sqrt2 functions in your global environment. Test my.sqrt by calling it with values that are both valid and invalid.

4. Open largest.R in RStudio and make my.largest a function in your global environment. Perform tests using column two or fourteen of the csv file *creditScreening*. This file contains 690 instances with missing items in the aforementioned rows. The file is used for several examples in later chapters.

5. Write the function everyOther which when given a list of numbers returns a list having every other item in the original list. For example, everyOther(c(1,2,3,4,5) returns (1,3,5).

6. Write the function *my.smallest* which when given one column of a data frame returns the smallest value in the column. Be sure to check for and print the number of NAs. Test your function using columns two and fourteen of the creditScreening dataset.

7. The remainder operator %% is used to determine if one number is a factor of another. For example, 2 is a factor of 6 as 6 %% 2 = 0. A perfect number is a number where the sum of its factors is equal to the number. Six is a perfect number as $1 + 2 + 3 = 6$. Write a function to list all perfect numbers between 1 and 1000. (Note: % / % gives the integer quotient of two numbers.)

8. Write a recursive version of my.largest. Use the technique given by my.SqrtR to initiate the recursion. Is it a good idea to implement my.largest recursively? Why or why not?

9. Use a while loop to write a function that finds the sum of the first 1000 even integers. Is it a good idea to implement this function recursively? Why or why not?

10. Write the function my.reverse that, given a vector, returns the items in reverse order. my.reverse(c(1,2,3)) returns 3 2 1.

11. The Euclidean algorithm provides an efficient way to find the greatest common divisor (GCD) of two positive integer numbers—the largest number that divides both numbers without a remainder. Write an iterative or recursive function to find the GCD. Here is the algorithm:

    a. Divide the smallest number into the largest number.

    b. If the remainder is 0, the divisor is the GCD and you are done.

    c. If there is a remainder, make the remainder the new smallest number and the divisor the new largest number. Repeat b.

Installed Packages and Functions

Package Name	Function(s)
base / stats	abs, as.integer, array, c, cat, cbind, data.frame, factor, for, function is.integer, is.na, is.numeric, list, matrix, mode, ncol, nrow, return, round, sapply sqrt, str, table, while
sqldf	sqldf

## CHAPTER 4

Preparing the Data

In This Chapter

- The KDD Process Model

- Relational Databases

- Data Preprocessing Techniques

- Data Transformation Methods

MANY OF THE PACKAGES available thru the Comprehensive R Archive Network (CRAN) repository include datasets which have been at least partially preprocessed. This allows us to concentrate on learning about the various machine learning tools without concerning ourselves with the preprocessing tasks needed to analyze real-world data. However, real data often contains missing values, noise, and requires one or several transformations before it is ready for the model building process. Therefore, before turning our attention to the machine learning algorithms presented in the next chapters, we examine solution strategies for the preprocessing issues often seen with real data.

In Section 4.1, we introduce a formal seven-step process model for knowledge discovery. In Sections 4.2–4.4, we concentrate on the steps involving the creation of initial target data, data preprocessing, and transformation as they are by far the most difficult and time-consuming parts of this process. When careful attention is paid to data preprocessing and data transformation, our chances of success in building useful machine learning models significantly increase.

4.1 A PROCESS MODEL FOR KNOWLEDGE DISCOVERY

Knowledge discovery in data (KDD) is an interactive, iterative procedure that attempts to extract implicit, previously unknown, and potentially useful knowledge from data. Several variations of the KDD process model exist. Variations describe the KDD process from 4 to as many as 12 steps. Although the number of steps may differ, most descriptions show consistency in content. Here is a brief description of a seven-step KDD process model:

1. *Goal identification.* The focus of this step is on understanding the domain being considered for knowledge discovery. We write a clear statement about what is to be accomplished. A hypothesis offering a likely or desired outcome can be stated.

2. *Creating a target dataset.* With the help of one or more human experts and knowledge discovery tools, we choose an initial set of data to be analyzed.

3. *Data preprocessing.* We use available resources to deal with noisy data. We decide what to do about missing data values and how to account for time-sequence information.

4. *Data transformation.* Attributes and instances are added and/or eliminated from the target data. We decide on methods to normalize, convert, and smooth data.

5. *Data mining.* A best model for representing the data is created by applying one or more machine learning algorithms.

6. *Interpretation and evaluation.* We examine the output from step 5 to determine if what has been discovered is both useful and interesting. Decisions are made about whether to repeat previous steps using new attributes and/or instances.

7. *Taking action.* If the discovered knowledge is deemed useful, the knowledge is incorporated and applied directly to appropriate problems.

As you work through the remaining sections of this chapter, we point out several R functions that implement the preprocessing techniques described here. These functions are used to help solve problems throughout the remaining chapters of your text. It is well worth your time to further investigate these functions prior to moving on to Chapters 5–12.

4.2 CREATING A TARGET DATASET

A viable set of resource data is of primary importance for any analytics project to succeed. Target data is commonly extracted from three primary sources—a data warehouse, one or more transactional databases, or one or several flat files. Many machine learning tools require input data to be in a flat file or spreadsheet format (i.e., R's data frame). If the original data is housed in a flat file, creating the initial target data is straightforward. Let's examine the other possibilities.

Database management systems (DBMS) store and manipulate transactional data. The computer programs in a DBMS are able to quickly update and retrieve information from a stored database. The data in a DBMS is often structured using the relational model. A *relational database* represents data as a collection of tables containing rows and columns. Each column of a table is known as an attribute, and each row of the table stores information about one data record. The individual rows are called *tuples*. All tuples in a relational table are uniquely identified by a combination of one or more table attributes.

A main goal of the relational model is to reduce data redundancy so as to allow for quick access to information in the database. A set of normal forms that discourage data redundancy define formatting rules for relational tables. If a relational table contains redundant data, the redundancy is removed by decomposing the table into two or more relational structures. In contrast, the goal of knowledge discovery is to uncover the inherent redundancy in data. Therefore, one or more relational join operations are usually required to restructure data into a form amenable for data mining.

To see this, consider the hypothetical credit card promotion database we defined in Table 2.2 of Chapter 2. Recall the table attributes: *income range, magazine promotion, watch promotion, life insurance promotion, credit card insurance, gender,* and *age.* The data in Table 2.2 is not a database at all but represents a flat file structure extracted from a database such as the one shown in Figure 4.1. The Acme credit card database contains tables about credit card billing information and orders, in addition to information about credit card promotions. The Promotion-C table creates two one-to-many relationships to resolve the single many-to-many relationship between Customer and Promotion. Therefore, the promotional information shown in Table 2.2 is housed in several relational tables within the database. The next section looks the work needed to extract information from the relational database of Figure 4.1 in order to create a data frame structure similar to Table 2.2.

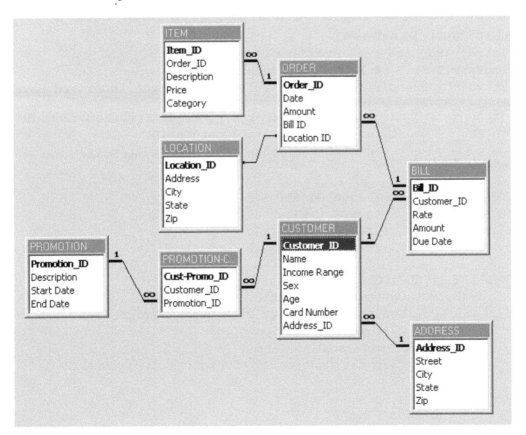

FIGURE 4.1 The Acme credit card database.

4.2.1 Interfacing R with the Relational Model

We have more than one option for connecting R to a relational database. For our example, we chose SQLite for the DBMS and R's DBI package for the database interface. We selected SQLite as it is highly portable, easy to install and use, provides the features of an SQL engine, and most importantly doesn't require a server! In addition, SQLite can be accessed using command line prompts or with a database browser.

The URL for SQLite's home page is https://sqlite.org/index.html. Click *download*, and then select the appropriate download link. For windows users, it is best to select the *sqlite tools* link under Precompiled Binaries for Windows. To avoid using the command line interface, you can download DB Browser—https://sqlitebrowser.org/dl/—which provides a nice interface that works on all major platforms.

We used SQLite to create a subset of the relational database shown in Figure 4.1. The database—CreditCardPromotion.db—is part of your supplementary materials for this chapter. Figure 4.2 displays the database structure as seen with DB Browser. Script 4.1 gives the statements for accessing the database. Prior to executing the script, you must visit the CRAN repository and install two interface packages—RSQLite and DBI. Let's take a look at the statements in Script 4.1.

Script 4.1 Creating a Data Frame with Data from a Relational Database

```
  library(RSQLite)# R interface to SQLite
> library(DBI)     # R database interface
> setwd("C:/Users/richa/desktop/sqlite")#Database is stored here

> dbCon <- dbConnect(SQLite(), dbname = "CreditCardPromotion.db")
> custab <- dbGetQuery(dbCon,"Select CustomerID,
+               IncomeRange,Gender,Age from Customer")
> custab

  CustomerID IncomeRange Gender Age
1          1       40-50K   Male  45
2          2       30-40K Female  40
3          3       40-50K Female  42

> ccpLife <- dbGetQuery(dbCon,
+      "Select Customer.CustomerID, IncomeRange,
+             Gender,Age, Status,Promotion_C.PromotionID
+      from   Customer, Promotion_C
+      where  PromotionID =10 and
+             Customer.CustomerID =Promotion_C.CustomerID
+             ")
> ccpLife

  CustomerID IncomeRange Gender Age Status PromotionID
1          1      40-50K   Male  45    Yes          10
```

| 2 | 2 | 30-40K | Female | 40 | Yes | 10 |
| 3 | 3 | 40-50K | Female | 42 | No | 10 |

```
> # Change column name from Status to LifeInsPromo
> colnames(ccpLife)[colnames(ccpLife)=="status"]<-"LifeInsPromo"
> ccpLife
```

	CustomerID	IncomeRange	Gender	Age	LifeInsPromo	PromotionID
1	1	40-50K	Male	45	Yes	10
2	2	30-40K	Female	40	Yes	10
3	3	40-50K	Female	42	No	10

```
> dbDisconnect(dbCon)
```

The first two statements load the installed libraries mentioned above. Modify the *setwd* statement to match the location of your copy of the database. The *dbConnect* function makes the connection to the database. The first *Select* statement obtains information from the *Customer* table. Notice that the database contains three customers.

You can see how the promotion IDs correspond to promotion names with a click within DB Browser on *browse data* or by submitting the following query in your console window:

```
dbGetQuery(dbCon, "Select * from Promotion")
```

	Description	PromotionID
1	Magazine Promotion	20
2	Watch Promotion	30
3	Life Insurance Promotion	10

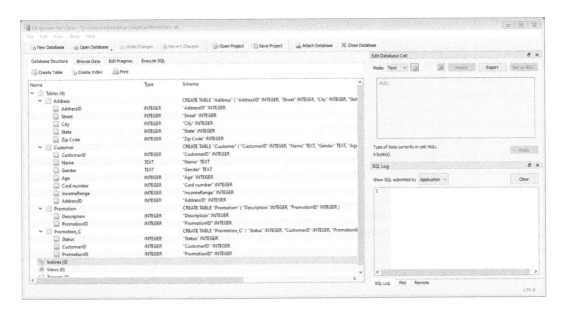

FIGURE 4.2   The CreditCardPromotion database.

Replace *Promotion* with *Promotion_C* to see how all three customers reacted to the three promotions.

The next query is of more interest as its purpose is to create the *Life Insurance Promotion* column shown in Table 2.2. To achieve this, the query joins the *Customer* and *Promotion_C* tables. The join creates a new table that combines the *Status* and *PromotionID* columns from Promotion_C with the information in the *Customer* table. Lastly, we use the *colnames* function to change *Status* to *LifeInsPromo* giving the desired result. Adding the magazine and watch promotions to ccpLife is left as an exercise. As you can see, when the data is stored in a relational database, the task of extracting target data can be a challenge.

4.2.2 Additional Sources for Target Data

The possibility also exists for extracting data from multiple databases. If target data is to be taken from more than one source, the transfer process can be tedious. Consider a simple example where one operational database stores customer gender with the coding *male = 1*, *female = 2*. A second database stores the gender coding as *male = M* and *female = F*. The coding for *male* and *female* must be consistent throughout all records in the target data or the data will be of little use. The process of promoting this consistency when transporting the data is a form of *data transformation*. Other types of data transformations are discussed in Section 4.4.

A third possibility for harvesting target data is the data warehouse. The data warehouse is a historical database designed for decision support rather than transaction processing (Kimball et al., 1998). Thus, only data useful for decision support is extracted from the operational environment and entered into the warehouse database. Data transfer from the operational database to the warehouse is an ongoing process usually accomplished on a daily basis after the close of the regular business day.

A fourth scenario is when data requires a distributed environment supported by a cluster of servers. With distributed data, the KDD process model must be supplemented with the added complexities of data distribution and solution aggregation. Lastly, a special case exists with streaming data where real-time analysis makes data preprocessing at best extremely difficult.

4.3 DATA PREPROCESSING

Most data preprocessing is in the form of data cleaning, which involves accounting for noise and dealing with missing information. Ideally, the majority of data preprocessing takes place before data is permanently stored in a structure such as a data warehouse.

4.3.1 Noisy Data

Noise represents random error in attribute values. In very large datasets, noise can come in many shapes and forms. Common concerns with noisy data include the following:

- How do we find duplicate records?

- How can we locate incorrect attribute values?

- What data smoothing operations should be applied to our data?

- How can we find and process outliers?

- How do we deal with missing data items?

4.3.1.1 Duplicate Records

Suppose a certain weekly publication has 100,000 subscribers and 0.1% of all mailing list entries have erroneous dual listings under a variation of the same name (e.g., Jon Doe and John Doe). Therefore, 100 extra publications are processed and mailed each week. At a processing and mailing cost of $2.00 for each publication, the company spends over $10,000 each year in unwarranted costs.

4.3.1.2 Incorrect Attribute Values

Finding errors in categorical data presents a problem in large datasets. Most data mining tools offer a summary of frequency values for categorical attributes. We should consider attribute values having frequency counts near 0 as errors.

A numeric value of 0 for an attribute such as blood pressure or weight is an obvious error. Such errors often occur when data is missing and default values are assigned to fill in for missing items. In some cases, such errors can be seen by examining class mean and standard deviation scores. However, if the dataset is large and only a few incorrect values exist, finding such errors can be difficult.

4.3.1.3 Data Smoothing

Data smoothing is both a data cleaning and data transformation process. Several data smoothing techniques attempt to reduce the number of values for a numeric attribute. Some classifiers, such as neural networks, use functions that perform data smoothing during the classification process. When data smoothing is performed during classification, the data smoothing is said to be internal. External data smoothing takes place prior to classification. Rounding and computing mean values are two simple external data smoothing techniques. Mean value smoothing is appropriate when we wish to use a classifier that does not support numerical data and would like to retain coarse information about numerical attribute values. In this case, all numerical attribute values are replaced by a corresponding class mean.

Another common data smoothing technique attempts to find atypical (outlier) instances in the data. Outliers often represent errors in the data whereby the items should be corrected or removed. For instance, a credit card application where applicant age is given as -21 is clearly incorrect. In other cases, it may be counterproductive to remove outliers. For example, in credit card fraud detection, the outliers are those items we are most interested in finding.

Unsupervised outlier detection methods often make the assumption that ordinary instances will cluster together. If definite patterns in the data do not exist, unsupervised techniques will flag an undue number of ordinary instances as outliers.

4.3.2 Preprocessing with R

Let's use the *creditScreening.csv* dataset to illustrate some basic preprocessing techniques employing R functions. This dataset is a viable choice as it contains numeric, categorical, and missing data items. The dataset includes information about 690 individuals who applied for a credit card. The data has 15 input attributes and one output attribute indicating whether an individual credit card application was accepted (+) or rejected (–). Privacy issues prevent knowledge of the semantic meaning of the input attributes. A more complete description of the dataset is given in Section 5.4 of Chapter 5. The output of the *str* function for the first three attributes and the class attribute gives the following:

```
> str(creditScreening)
'data.frame': 690 obs. of 16 variables:
 $ one      : Factor w/ 3 levels "?","a","b": 3 2 2 3 3 3 3 2 3 3
 ...
 $ two      : num  30.8 58.7 24.5 27.8 20.2 ...
 $ three    : num  0 4.46 0.5 1.54 5.62 ...
 $ class    : Factor w/ 2 levels "-","+": 2 2 2 2 2 2 2 2 2 2 ...
```

The *str* function tells us the file has been imported as a data frame. Also, attribute *one* lists one of its values as "?" indicating a likely unknown attribute value. The subset function can give us the rows of those instances with attribute *one* having a value of "?". Here are the first four such instances obtained with the subset function.

```
subset(creditScreening, one=="?")
```

	one	two	three	four	five	six	seven	eight	nine	ten	eleven	twelve
218	?	40.83	3.500	u	g	i	bb	0.500	f	f	0	f
237	?	32.25	1.500	u	g	c	v	0.250	f	f	0	t
265	?	28.17	0.585	u	g	aa	v	0.040	f	f	0	f
344	?	29.75	0.665	u	g	w	v	0.250	f	f	0	t

......

The summary function provides information about missing items for both categorical and numeric attributes. Here is summary information for attributes *one* through *six*:

```
summary(creditScreening)
```

one	two	three	four	five	six
?: 12	Min. :13.75	Min. : 0.000	: 6	: 6	c :137
a:210	1st Qu.:22.60	1st Qu.: 1.000	l: 2	g :519	q : 78
b:468	Median :28.46	Median : 2.750	u:519	gg: 2	w : 64
	Mean :31.57	Mean : 4.759	y:163	p :163	i : 59
	3rd Qu.:38.23	3rd Qu.: 7.207			aa : 54
	Max. :80.25	Max. :28.000			ff : 53
	NA's :12				(Other):245

The summary information above tells us attribute *two* contains 12 NA's. We also see the "?" for attribute *one*. The *unique* function lists the unique values found in a column of data. Attribute *six* will likely be of little predictive value as it has 16 unique values:

```
length(unique(creditScreening[,6]))
[1] 10
```

Attributes *nine*, *ten*, and *twelve* (not shown) are more likely to be useful as they represent even distributions of true and false values.

4.3.3 Detecting Outliers

Graphical approaches are often employed for outlier detection. Figure 4.3 shows a histogram of creditScreening$fourteen. The histogram shows that the greatest majority of values lie well below 1000. However, the histogram clearly indicates a very small set of values in the 1700–2000 range. Given the summary statement tells us attribute fourteen has 14 missing items, these outliers are likely errors in the data.

There are also functions for detecting outliers. We will investigate the *outlierTest* function available with the *car* package in Chapter 5. Since outlier detection operators do not take attribute significance into account, the outliers detected by an operator may not be useful. When using machine learning methods such as neural networks that do not have attribute selection built into the modeling process, it is best to first apply an attribute selection technique to the data prior to attempting outlier detection.

4.3.4 Missing Data

Imputation is a general term used for replacing missing data with substituted values. In most cases, missing attribute values indicate lost information. For example, a missing value for the attribute *age* certainly indicates a data item that exists but is unaccounted for. However, a missing value for *salary* may be taken as an unentered data item, but it could also indicate an individual who is unemployed. Some machine learning techniques are able to deal directly with missing values. However, many methods require all attributes to contain a value.

The following are possible options for dealing with missing data *before* the data is presented to a learning algorithm.

- *Discard records with missing values.* This method is most appropriate when a small percent of the total number of instances contain missing data and we can be certain that missing values do indeed represent lost information.

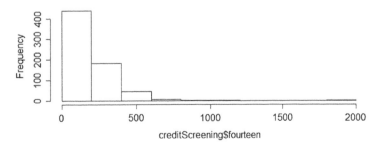

FIGURE 4.3   A histogram of attribute creditScreening$fourteen.

- *For real-valued data, replace missing values with the class mean.* In most cases, this is a reasonable approach for numerical attributes. Options such as replacing missing numeric data with a zero or some arbitrarily large or small value is generally a poor choice.

- *Replace missing attribute values with the values found within other highly similar instances.* This technique is appropriate for either categorical or numeric attributes.

Some machine learning techniques allow instances to contain missing values. Here are three ways that machine learning techniques deal with missing data while learning:

1. *Ignore missing values.* Several machine learning algorithms, including neural networks (Chapter 8) and Bayes classifier (Chapter 5), use this approach.

2. *Treat missing values as equal comparisons.* This approach is dangerous with very noisy data in that dissimilar instances may appear to be very much alike.

3. *Treat missing values as unequal comparisons.* This is a pessimistic approach but may be appropriate. Two similar instances containing several missing values will appear dissimilar.

Finally, a knowledge-based approach for resolving missing information uses supervised learning to determine likely values for missing data. When the missing attribute is categorical, we designate the attribute with missing values as an output attribute. The instances with known values for the attribute are used to build a classification model. The created model is then summoned to classify the instances with missing values. For numerical data, we can use a regression technique or a neural network and apply the same strategy.

Two functions located in *functions.zip* written for your text are designed to help detect and replace missing items. When given a data frame, *removeNAS* lists all instances having at least one NA. It returns a data frame with all instances having missing numeric data removed. The second function, *RepVal*, replaces any item in a given column of a data frame with a specified value. For example, the call

```
y<-repVal(creditScreening, 1, '?','a')
```

creates a new data frame y where all '?' values in column one are replaced with an 'a'.

4.4 DATA TRANSFORMATION

Data transformation can take many forms and is necessary for a variety of reasons. We offer a description of some familiar data transformations in the following sections.

4.4.1 Data Normalization

A common data transformation involves changing numeric values so they fall within a specified range. Classifiers such as neural networks do better with numerical data scaled to a range between 0 and 1. Normalization is particularly appealing with distance-based classifiers, because by normalizing attribute values, attributes with a wide range of values are less likely to outweigh attributes with smaller initial ranges. Four common normalization methods include:

- *Decimal scaling.* Decimal scaling divides each numerical value by the same power of 10. For example, if we know the values for an attribute range between –1000 and 1000, we can change the range to –1 and 1 by dividing each value by 1000.

- *Min-Max normalization.* Min-Max is an appropriate technique when minimum and maximum values for an attribute are known. The formula is

$$NewValue = \frac{originalValue - oldMin(NewMax - newMin) + newMin}{oldMax - oldMin}$$

where *oldMax* and *oldMin* represent the original maximum and minimum values for the attribute in question. *NewMax* and *newMin* specify the new maximum and minimum values. *NewValue* represents the transformation of *originalValue*. This transformation is particularly useful with neural networks where the desired range is [0,1]. In this case, the formula simplifies to

$$NewValue = \frac{originalValue - oldMin}{oldMax - oldMin}$$

- *Normalization using Z-scores.* Z-score normalization converts a value to a standard score by subtracting the attribute mean (μ) from the value and dividing by the attribute standard deviation (σ). Specifically,

$$newValue = \frac{originalValue - \mu}{\sigma}$$

This technique is particularly useful when maximum and minimum values are not known.

- *Logarithmic normalization.* The base *b* logarithm of a number *n* is the exponent to which *b* must be raised to equal *n*. For example, the base 2 logarithm of 64 is 6 because $2^6 = 64$. Replacing a set of values with their logarithms has the effect of scaling the range of values without loss of information.

R's generic *scale* function scales a column of numeric data to a mean of 0 and a standard deviation of 1. Execute the following statements to get a clear picture of how the *scale* function normalizes *ccpromo$Age*.

```
my.sc <- scale(ccpromo$Age)
str(my.sc)
my.sc <- data.frame(col1=my.sc)
```

```
str(my.sc)
my.sc<-as.numeric(my.sc$col1)
my.sc
sd(my.sc)
mean(my.sc)
```

4.4.2 Data Type Conversion

Many machine learning tools, including neural networks and some statistical methods, cannot process categorical data. Therefore, converting categorical data to a numeric equivalent is a common data transformation.

4.4.3 Attribute and Instance Selection

Classifiers such as decision trees with built-in attribute selection are less likely to suffer from the effects of datasets containing attributes of little predictive value. Unfortunately, many machine learning algorithms such as neural networks and nearest neighbor classifiers are unable to differentiate between relevant and irrelevant attributes. This is a problem, as these algorithms do not generally perform well with data containing a wealth of attributes that are not predictive of class membership. Furthermore, it has been shown that the number of training instances needed to build accurate supervised models is directly affected by the number of irrelevant attributes in the data. To overcome these problems, we must make decisions about which attributes and instances to use when building our models. The following is a possible algorithm to help us with attribute selection:

1. Given N attributes, generate the set S of all possible attribute combinations.
2. Remove the first attribute combination from set S and generate a machine learning model M using these attributes.
3. Measure the goodness of model M.
4. Until S is empty
   a. Remove the next attribute combination from S and build a model using the next attribute combination in S.
   b. Compare the goodness of the new model with the saved model M. Call the better model M and save this model as the best model.
5. Model M is the model of choice.

This algorithm will surely give us a best result. The problem with the algorithm lies in its complexity. If we have a total of n attributes, the total number of attribute combinations is $2^n - 1$. The task of generating and testing all possible models for any dataset containing more than a few attributes is not possible. Let's investigate a few techniques we can apply.

4.4.3.1 Wrapper and Filtering Techniques

Attribute selection methods are generally divided into filtering and wrapper techniques. *Filtering* methods select attributes based on some measure of quality independent of the algorithm used to build a final model. With *wrapper* techniques, attribute goodness is measured in the context of the learning algorithm used to build the final model. That is, the algorithm is wrapped into the attribute selection process. You will see several examples of both methods in the chapters that follow.

4.4.3.2 More Attribute Selection Techniques

In addition to applying wrapper and filtering techniques, several other steps can be taken to help determine which attributes to eliminate from consideration:

1. Highly correlated input attributes are redundant. Most machine learning tools build better models when only one attribute from a set of highly correlated attributes is designated as an input value. Methods for detecting correlated attributes include graphical techniques as well as quantitative measures.

2. *Principal component analysis* is a statistical technique that looks for and removes possible correlational redundancy within the data by replacing the original attributes with a smaller set of artificial values.

3. A *self-organizing map* or SOM is a neural network trained with unsupervised learning that can be used for attribute reduction. SOMs are described in Chapter 8 when we focus on neural network models.

4.4.3.3 Creating Attributes

Attributes of little predictive power can sometimes be combined with other attributes to form new attributes with a high degree of predictive capability. As an example, consider a database consisting of data about stocks. Conceivable attributes include current stock price, 12-month price range, growth rate, earnings, market capitalization, company sector, and the like. The attributes price and earnings are of some predictive value in determining a future target price. However, the ratio of price to earnings (P/E ratio) is known to be more useful. A second created attribute likely to effectively predict a future stock price is the stock P/E ratio divided by the company growth rate. Here are a few transformations commonly applied to create new attributes:

- Create a new attribute where each value represents a ratio of the value of one attribute divided by the value of a second attribute.

- Create a new attribute whose values are differences between the values of two existing attributes.

- Create a new attribute with values computed as the percent increase or percent decrease of two current attributes. Given two values v_1 and v_2 with $v_1 < v_2$, the percent increase of v_2 with respect to v_1 is computed as

$$Percent\ Increase(v_2, v_1) = \frac{v_2 - v_1}{v_1}$$

If $v_1 > v_2$, we subtract v_2 from v_1 and divide by v_1, giving a percent decrease of v_2 with respect to v_1.

New attributes representing differences and percent increases or decreases are particularly useful with time series analysis. *Time series analysis* models changes in behavior over a time interval. For this reason, attributes created by computing differences between one time interval and the next time interval are important.

4.4.4 Creating Training and Test Set Data

Once we have our data preprocessed and transformed, our final step prior to model building is the selection of training and test data. A common scenario for supervised learning is to start by randomizing the data. This is followed by selecting 2/3 of the data for model building and the remaining 1/3 for testing. This is illustrated below using the creditScreening data. Let's take a look at each statement.

```
> set.seed(1000)
> # Randomize and split the data for 2/3 training, 1/3 testing
> credit.data <- creditScreening
> index <- sample(1:nrow(credit.data), 2/3*nrow(credit.data))
> credit.train <- credit.data[index,]
> credit.test <-  credit.data[-index,]
```

The *set.seed* function generates an initial starting point for the pseudo-random number generator. Using the same seed each time makes our experiments reproducible. Next, a copy of the creditScreening data frame is made for the experiment. The *sample* function is given two arguments. The first argument is the total number of rows in the data frame—690 in this case. The second argument specifies the size of the sample. In this case, index will contain a list of 460 randomized positive integer values between 1 and 690.

The statement `credit.data[index,]` copies all of the instances with row numbers specified by index into `credit.train`. Finally, `credit.data[-index,]` specifies `credit.test` is to receive a copy of the remaining 1/3 of the instances. If data normalization is needed, the *scale* function would precede sampling.

We have provided but one method for randomizing and splitting data for training and testing. Countless other techniques can be used to achieve the same result. The next section overviews an alternative to the training–testing scenario when data is lacking.

4.4.5 Cross Validation and Bootstrapping

If ample test data are not available, we can apply a technique known as *cross validation*. With this method, all available data are partitioned into n fixed-size units. $n-1$ of the units are used for training, whereas the nth unit is the test set. This process is repeated until each of the fixed-size units has been used as test data. Model test set correctness is computed as the average accuracy realized from the n training–testing trials. Experimental results have shown a value of 10 for n to be maximal in most situations. Several applications of cross validation to the data can help ensure an equal distribution of classes within the training and test datasets.

Bootstrapping is an alternative to cross validation. With bootstrapping, we allow the training set selection process to choose the same training instance several times. This happens by placing each training instance back into the data pool after it has been selected for training. It can be shown mathematically that if a dataset containing n instances is sampled n times using the bootstrap technique, the training set will contain approximately two-thirds of the n instances. This leaves one-third of the instances for testing.

4.4.6 Large-Sized Data

While traditional machine learning algorithms assume the entire dataset resides in memory, current datasets are often too large to satisfy this requirement. One possible way to deal with this problem is to process as much of the data as possible while repeatedly retrieving the remaining data from a secondary storage device. Clearly, this solution is not reasonable given the inefficiency of secondary storage data retrieval. Another possibility is to employ a distributed environment where data can be divided among several processors. When this is not feasible, we must limit our list of plausible algorithm choices to those that exhibit the property of scalability.

An algorithm is said to be *scalable* if given a fixed amount of memory, its runtime increases linearly with the number of records in the dataset. The simplest approach to scalability is sampling. With this technique, models are built and tested using a subset of the data that can be efficiently processed in memory. This method works well with supervised learning and unsupervised clustering when the data contains a minimal number of outliers. However, it would be difficult to put our trust in this method as a general approach for handling large-sized data.

Some traditional algorithms such as Naïve Bayes classifier (Chapter 5) are scalable. Cobweb and Classit (Chapter 11) are two scalable unsupervised clustering techniques as data is processed and discarded incrementally.

4.5 CHAPTER SUMMARY

Knowledge discovery can be modeled as a seven-step process that includes goal identification, target data creation, data preprocessing, data transformation, data mining, result interpretation and evaluation, and knowledge application. A clear statement about what is to be accomplished is a good starting point for successful knowledge discovery. Creating a target dataset often involves extracting data from a warehouse, a transactional database, or a distributed environment. Transactional databases do not store redundant data, as they are modeled to quickly update and retrieve information. Because of this, the structure of the data in a transactional database must be modified before data mining can be applied.

Prior to exercising a machine learning tool, the gathered data is preprocessed to remove noise. Missing data is of particular concern because many algorithms are unable to process missing items. In addition to data preprocessing, data transformation techniques can be applied before model building takes place. Data transformation methods such as data normalization and attribute creation or elimination are often necessary for a best result.

4.6 KEY TERMS

- *Attribute filtering.* Attribute filtering methods select attributes based on some measure of quality independent of the algorithm that will be used to build a final model.

- *Bootstrapping.* Allowing instances to appear more than once in a training set.

- *Cross validation.* Partitioning a dataset into *n* fixed-size units. *n* – 1 units are used for training and the *n*th unit is used as a test set. This process is repeated until each of the fixed-size units has been used as test data. Model test set correctness is computed as the average accuracy realized from the *n* training–testing trials.

- *Data normalization.* A data transformation where numeric values are modified to fall within a specified range.

- *Data preprocessing.* The step of the KDD process that deals with noisy and missing data.

- *Data transformation.* The step of the KDD process that deals with data normalization and conversion as well as the addition and/or elimination of attributes.

- *Decimal scaling.* A data transformation technique for a numeric attribute where each value is divided by the same power of 10.

- *Logarithmic normalization.* A data transformation method for a numeric attribute where each numeric value is replaced by its base b logarithm.

- *Min-Max normalization.* A data transformation method that is used to transform a set of numeric attribute values so they fall within a specified numeric range.

- *Noise.* Random error in data.

- *Outlier.* An instance that by some measure deviates significantly from other instances.

- *Relational database.* A database where data is represented as a collection of tables containing rows and columns. Each column of the table is known as an attribute, and each row of the table stores information about one data record.

- *Time series analysis.* Any technique that models changes in behavior over a period of time.

- *Tuple.* An individual row of a table in a relational database.

- *Wrapper technique.* An attribute selection method that bases attribute goodness in the context of the learning algorithm used to build the final model.

- *Z-score normalization.* A data normalization technique for a numeric attribute where each numeric value is replaced by its standardized difference from the mean.

EXERCISES

Review Questions

1. Differentiate between the following terms:

   a. Data cleaning and data transformation

   b. Internal and external data smoothing

   c. Decimal scaling and Z-score normalization

   d. Filter and wrapper attribute selection

2. In Section 4.4, you learned about basic methods machine learning algorithms use to deal with missing data while learning. Decide which technique is best for the following problems. Explain each choice.

   a. A model designed to accept or reject credit card applications.

   b. A model for determining who should receive a promotional flyer in the mail.

   c. A model designed to determine those individuals likely to develop colon cancer.

   d. A model to decide whether to drill for oil in a certain region.

   e. A model for approving or rejecting candidates applying to refinance their home.

Experimenting with R

1. Consider Script 4.1 Copy the script and add two SQL Select clauses to the new script that will create the columns of yes and no responses for *magazine promotion* and *watch promotion*. Use *cbind* to add each column to ccpLife. Also, be sure to remove the PromotionID column from the final table. Your final output will show a single table with the headings: CustomerID, IncomeRange, Gender, Age, LifeInsPromo, MagazinePromo, and WatchPromo.

Computational Questions

1. Set up a general formula for a Min-Max normalization as it would be applied to the attribute *age* for the data in Table 2.2. Transform the data so the new minimum value is 0 and the new maximum value is 1. Apply the formula to determine a transformed value for *age* = 35.

2. Answer the following questions about percent increase and percent decrease.

    a. The price of a certain stock increases from $25.00 to $40.00. Compute the percent increase in the stock price.

    b. The original price of the stock is $40.00. The price decreases by 50%. What is the current stock price?

3. You are to apply a base 2 logarithmic normalization to a certain numeric attribute whose current range of values falls between 2300 and 10,000. What will be the new range of values for the attribute once the normalization has been completed?

4. Apply a base 10 logarithmic normalization to the values for attribute *age* in Table 2.2. Use a table to list the original values as well as the transformed values.

Installed Packages and Functions

Package Name	Function(s)
base / stats	*abs, c, cat, cbind, colnames, data.frame, factor, length, library, ncol, nrow, return, round, sample, scale, set.seed, setwd, str, subset, summary, table, unique*
car	*outlierTest*
DBI & RSQLite	*dbConnect, dbDisconnect, dbGetQuery*

## CHAPTER 5

Supervised Statistical Techniques

In This Chapter

- Linear Regression

- Logistic Regression

- Naïve Bayes Classifier

- Model Evaluation

MATHEMATICS AND STATISTICS SET much of the ground work for the field of machine learning. As such, a good starting point for our discussion of machine learning techniques is an examination of supervised statistical methods. These methods are labeled *statistical* in that they make certain assumptions about the data. If these assumptions are violated, the significance tests performed on the results may be inaccurate. With these techniques, input attributes are frequently referred to as *independent variables* or *features*, and output attributes are described as *response* or *dependent variables*.

In Section 5.1, we start by looking at simple linear regression where a single input attribute determines a numeric outcome. The focus of Section 5.2 is multiple linear regression where several input attributes establish a numeric result. In Section 5.3, we discuss logistic regression and how it is applied to build supervised learner models for datasets with a binary outcome. In Section 5.4, you learn how the Bayes classifier builds supervised models for both categorical and real-valued input data. Let's begin!

5.1 SIMPLE LINEAR REGRESSION

Statistical *regression* is a supervised technique that generalizes a set of numeric data by creating a mathematical equation relating one or more input attributes to a single output attribute. With *linear regression*, we attempt to model the variation in a dependent variable as a linear combination of one or more independent variables. A linear regression equation is of the form:

$$f(x_1, x_2, x_3 \ldots x_n) = a_1 x_1 + a_2 x_2 + a_3 x_3 + \cdots + a_n x_n + c \tag{5.1}$$

where $x_1, x_2, x_3, \ldots, x_n$ are independent variables and $a_1, a_2, a_3, \ldots, a_n$ and c are constants. $f(x_1, x_2, x_3, \ldots, x_n)$ represents the dependent variable and is often shown simply as y. In general, linear regression is a favorite among statisticians and tends to work well when the relationship between the dependent and independent variables is nearly linear.

The simplest form of the linear regression equation allows but a single independent variable as the predictor of the dependent variable. This type of regression is appropriately named *simple linear regression*. The regression equation is written in *slope–intercept form*. Specifically,

$$y = ax + b \tag{5.2}$$

where x is the independent variable and y depends on x. The constants a and b are computed via supervised learning by applying a statistical criterion to a dataset of known values for x and y. The graph of Equation 5.2 is a straight line with slope a and y intercept b.

A common statistical measure used to compute a and b is the *least-squares criterion*. The least-squares criterion minimizes the sum of squared differences between actual and predicted output values. Deriving a and b via the least-squares method requires a knowledge of differential calculus. Therefore, we simply state the formulas for computing a and b. For a total of n instances, we have:

$$b = \frac{\sum xy}{\sum x^2} \quad a = \frac{\sum y}{\sum n} - \frac{b \sum y}{n} \tag{5.3}$$

With these basics in hand, let's create a regression model using a subset of the data described in the box titled "The Gamma-Ray Burst Dataset". The dataset is interesting as gamma-ray bursts are the most powerful singular event in our universe and there is considerable disagreement as to the number of clusters present in the data (Mukherjee et al., 1998). Our example is designed to illustrate simple linear regression. Therefore, we must limit our experiment to two of the seven attributes within the data. Let's examine the degree of linear relationship between the two measures of burst length—*t90* and *t50*.

GAMMA-RAY BURST DATASET

Gamma-ray bursts are brief gamma-ray flashes with origins outside of our solar system. More than 1000 such events have been recorded. The gamma-ray burst data in this dataset are from the BATSE 4B catalog. The bursts in the BATSE 4B catalog were observed by the Burst And Transient Source Experiment (BATSE) aboard NASA's Compton Gamma-Ray Observatory between April 1991 and March 1993. Although many attributes have been measured for these bursts, the dataset is limited to seven attributes. Attribute *burst* gives the assigned burst number. All other attributes have been preprocessed by applying a logarithmic normalization. Normalized attributes *t90* and *t50* measure burst duration (burst length), *p256* and *fluence* measure burst brightness, and *hr321* and *hr32* measure burst hardness.

This dataset is interesting as it allows astronomers to develop and test various hypotheses about the nature of gamma-ray bursts. In doing so, astronomers have an opportunity to learn more about the structure of the universe. Also, the raw gamma-ray burst data had to be preprocessed and transformed several times before a set of significant attributes was developed. The dataset clearly demonstrates the importance of data preprocessing and data transformation. The data is strictly real-valued and listed under the name Grb4u.csv. If you would like more information about the BATSE project, visit the Web site at https://gammaray. nsstc.nasa.gov/batse/instrument/batse.html

Import the dataset *Grb4u.csv* and load *Script 5.1 Simple Linear Regression*—both part of the supplementary materials for the text—into your RStudio editor. The script contains the code to perform a simple linear regression using *t90* as the response variable and *t50* as the lone input variable. An edited version—some text has been removed to conserve space—of Script 5.1 is given below. Our goal is to examine the relationship between these two measures of gamma-ray burst length. You can execute your script line by line either by typing ctrl+enter in the console or by using your editor's *run* icon.

Script 5.1 Simple Linear Regression

```
> round(head(Grb4u),3)
   burst    p256      fl   hr32 hr321    t50    t90
1   1700   0.006  -5.705 0.514 0.280 0.730 1.376
2   4939   0.119  -5.580 0.469 0.219 0.850 1.302
3    606   0.022  -5.504 0.431 0.195 0.933 1.317

> grb <- Grb4u[c(6,7)]
> summary(grb)

      t50                 t90
 Min.   :-1.9208   Min.   :-1.6198
 Max.   : 2.6830   Max.   : 2.8285

> # Correlation(pearson)
> round(cor(grb$t90,grb$t50),3)
[1] 0.975

> # Create the regression model and Plot the data
> my.slr <- lm(t90 ~ t50, data =grb)

> plot(grb$t50,grb$t90,xlab="t50",
+        ylab="t90",main="Gamma-ray burst data t50 by t90")

> abline(my.slr) #Adds a line of best model fit

> summary(my.slr) # Analyze the results

Call:
lm(formula = t90 ~ t50, data = grb)
```

```
Residuals:
    Min        1Q    Median        3Q       Max
-0.44602  -0.13785  -0.02365   0.09597   1.45865

Coefficients:
             Estimate Std. Error t value Pr(>|t|)
(Intercept)  0.428164   0.007155   59.84   <2e-16 ***
grb$t50      0.985041   0.006603  149.18   <2e-16 ***
---
Signif. codes:  0 '***' 0.001 '**' 0.01 '*' 0.05 '.' 0.1 ' ' 1

Residual standard error: 0.2134 on 1177 degrees of freedom
Multiple R-squared:  0.9498,   Adjusted R-squared:  0.9497
F-statistic: 2.226e+04 on 1 and 1177 DF, p-value: < 2.2e-16

> head(round(fitted(my.slr),3))
    1     2     3     4     5     6
1.148 1.265 1.347 1.282 1.505 1.351

> head(round(residuals(my.slr),3))
     1      2      3      4      5      6
 0.228  0.037 -0.031  0.058 -0.017 -0.013
```

The first executable line in the script prints the first six instances in the dataset. The *round* function limits the number of places behind the decimal. The variable *grb* is created by extracting *t50* and *t90* from the original data. The *summary* function displays the range for *t90* and *t50*. The *cor* function returns a value above 0.95 telling us that *t50* and *t90* are highly correlated. This strongly implies the relationship between *t90* and *t50* is linear.

The R function for obtaining a linear regression model is *lm( )*. For our example, we have:

```
> my.slr <- lm(t90 ~ t50, data =grb)
```

The object *my.slr* stores information about the created model. The statement within the parentheses is given in two parts. The leftmost component is the regression formula where a tilde follows output attribute *t90*. Next comes the list of input attributes which in our case is *t50*. The name of the dataset follows the comma. Several equivalent parenthetical forms of this equation exist. Two of note include:

- (*grb$t90 ~ grb$t50, data = grb*)

- (*t90 ~ ., data= grb*)

In the first example, we see the variable *t90* is prefixed by the file name. Here, specifying the name of the dataset is not needed as the dataset reference is resolved by *data=grb*. The period-comma format given in the second example without an attribute reference implies all input attributes are to be used.

A + sign between input attributes is used to designate a subset of all available attributes. For example,

(t90 ~ t50 + fl + hr32, data = Grb4u)

states that *t50*, *fl*, and *hr32* are to be used to create a multiple linear regression model. This format may seem a bit cryptic at first but you will get used to it!

Additional information about *lm* can be accessed with *help(lm)* or by typing *?lm*.

The *plot* and *abline* functions give the graph displayed in Figure 5.1 The figure clearly shows the linear relationship between *t50* and *t90*. The grb$ can be avoided in the plot statement by using the *attach* function.

Parts of the output of the *summary* function are given in terms of probabilities and significance levels—topics detailed in Chapter 9. Here is a brief overview of the basic statistical concepts needed to better understand the output.

A *significance level* is the probability at which we are willing to reject that the result of an experiment is duev purely to chance. The significance level can also be defined as the probability of incorrectly rejecting a purely chance event. The significance level is usually referenced by the letter *p* and is chosen by the researcher prior to the beginning an experiment. Common choices for *p* are $p \leq 0.05$ or $p \leq 0.01$. For example, suppose we apply two different machine learning models (*A* and *B*) to a test dataset to help determine which model is a better choice for general use. We decide to use $p \leq 0.05$ as the significance level. Suppose the result of our experiment shows model *B* performs better than model *A* with $p = 0.03$. Given this result, we can be 95% confident that *A* is a better model choice. Alternatively, if $p \leq 0.01$ was chosen as the significance level, our conclusion would be that there is no significant difference in performance between the two models. Two quantitative measures commonly used to compute significance levels are the *t* and *F* statistics both of which are discussed in Chapter 9.

summary(my.slr) offers several items of interest. We first see the call to *lm*. Next, we have information about residuals. A *residual* represents an error and is computed as the difference between actual and computed output.

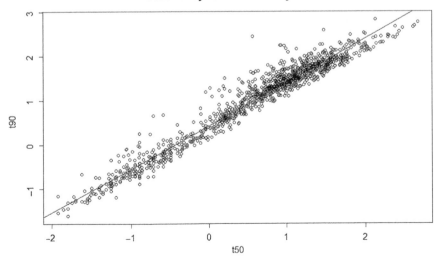

FIGURE 5.1   Gamma-ray burst data t50 by t90.

The *Coefficients* table gives us what we need to specify the regression equation. Specifically,

$$t90 = 0.985t50 + 0.428$$

Values under the heading *t value* are *t* scores computed as *Estimate* divided by *Std. Error*. Pr(>| *t* |) is the significance level at which we can be certain that the *t* value for the tested coefficient is significantly different than zero. As you can see, the certainty that the intercept and the coefficient for *t50* is not zero lies somewhere between 99% and 100%.

Multiple R-squared (0.9498) can be thought of as the correlation between actual and predicted values for *t90*. Therefore, it measures the amount of variation in the response variable that can be explained by the input variables. The adjusted R-squared normalizes the multiple R-squared value by taking into account the number of instances and the number of variables being used. An R-squared value approaching 1 indicates that the regression equation closely fits the output data.

The *residual standard error*—shown as 0.21—is the average error in predicting output values. It is computed as the square root of the sum of squares of all residuals divided by the degrees of freedom. *Degrees of freedom* is defined as the number of values in the calculation of a statistic that are free to vary. To illustrate, if I tell you the sum of three integers is 24, you are free to choose any value for two of the integers. However, to meet the summation criterion, the third choice must be an integer that when added to the sum of the first two gives 24. Therefore, the computation has two degrees of freedom.

For our example, the degrees of freedom for the *residual standard error* must take into account the *squared sum of the residuals* and the *regression model*. For the squared sum of the residuals, the error contribution for 1178 of the instances is free to vary. The final instance is bound to values that must give the correct summation for all squared errors. The degrees of freedom for the regression model corresponds to the number of coefficients that have been estimated minus 1. Including the intercept, there are two coefficients, so the regression model has 2 − 1 = 1 degrees of freedom. Therefore, the degrees of freedom for the residual standard error is 1178 − 1 = 1177.

The degrees of freedom associated with a particular model can be obtained with the function *df.residual(model)*. Also, the residual sum of squares value used to compute the residual standard error is given by the function *aov(model)*. However, a word of caution is in order as these functions are data dependent and cannot be used when a model is applied to new data.

The *p-value* associated with the *F statistic* compares an intercept-only model to the specified model and tells us if at least one of the input variables determines the output variable beyond chance. Stated another way, the *p-value* is the probability of obtaining a value for *F* at least as extreme as the computed value by random error alone. The *p-value* of 2.2e-16 all but removes any possibility of a random error event.

Two additional functions of interest are the *fitted* function which prints out the computed values of *t90*, and the *residuals* function that displays the error between actual and computed *t90* values. The output of these functions for the first six instances is given in the script.

The totality of these results strongly indicate that our regression equation closely models *t90*. This is not surprising, as *t50* is a second measure of gamma-ray burst distance. However, we must keep in mind that these statistics are based on the data used to build the model. We have no information about how the equation will perform when presented with unseen data. Lastly, simple linear regression is both easy to understand and apply. However, the fact that it allows but a single independent variable makes the technique of limited use for most machine learning applications. Let's expand our horizons with multiple linear regression!

5.2 MULTIPLE LINEAR REGRESSION

The least-squares criterion is also used to create linear regression equations of more than two variables. With Equation 5.1 written as

$$y = \sum a_i x_i + c \qquad (5.4)$$

and i varying from 1 to n, the regression algorithm determines c and the a_i values so as to minimize—over all instances—the sum of squared differences (e_i) between actual (y_i) and predicted ($\hat{y}_i$) values. The goal is to minimize the following:

$$\sum (y_i - \hat{y}_i)^2 = \sum (e_i)^2 \qquad (5.5)$$

Let's continue our investigation of gamma-ray burst data with an example using multiple linear regression. Our goal is to develop a regression model able to determine $t90$, but this time using burst brightness and hardness as the input features.

5.2.1 Multiple Linear Regression: An Example

Script 5.2 lists the procedure for our example with multiple linear regression. The first statement loads the *car package*—available in the CRAN repository—as it contains the scatterplotMatrix function which allows us to graphically examine scatterplot relationships between variable pairs. Next, we use the head function to remind us of the attribute names. The *cor* function illustrates high correlations between *p256 & fl, hr321 & hr32,* and *t50 & t90*. We will eliminate one of each pair of features including t50 as Script 5.1 showed us that t50 is a viable substitute for t90.

Script 5.2 Multiple Linear Regression: Gamma-Ray Burst Data

```
> library(car)

> # PREPROCESSING

> round(head(Grb4u,2),3)

burst    p256      fl   hr32 hr321    t50    t90
1   1700   0.006 -5.705 0.514 0.280 0.730 1.376
2   4939   0.119 -5.580 0.469 0.219 0.850 1.302

> round(cor(Grb4u),3)
```

```
           p256     fl    hr32   hr321    t50     t90
p256     1.000   0.602   0.170   0.181   0.013   0.073
fl       0.602   1.000  -0.030  -0.042   0.643   0.683
hr32     0.170  -0.030   1.000   0.959  -0.387  -0.391
hr321    0.181  -0.042   0.959   1.000  -0.407  -0.411
t50      0.013   0.643  -0.387  -0.407   1.000   0.975
t90      0.073   0.683  -0.391  -0.411   0.975   1.000


> grb.data <- Grb4u[c(3,4,7)]

> # PLOT THE DATA
> scatterplotMatrix(my.data,main="gamma-ray burst data")
> densityPlot(grb$t90)


> # BUILD THE MODEL

> grb.mlr <- lm(t90 ~ ., data = grb.data)


> #ANALYZE THE RESULTS
> summary(grb.mlr)

Call:
lm(formula = t90 ~ ., data = grb.data)

Residuals:
    Min      1Q   Median      3Q     Max
-4.2490 -0.3634   0.0061  0.3716  2.1796

Coefficients:
            Estimate Std. Error t value Pr(>|t|)
(Intercept)  6.20572    0.13105   47.35   <2e-16 ***
fl           0.83185    0.02275   36.57   <2e-16 ***
hr32        -1.18955    0.05896  -20.17   <2e-16 ***
---
Signif. codes:  0 '***' 0.001 '**' 0.01 '*' 0.05 '.' 0.1 ' ' 1

Residual standard error: 0.5997 on 1176 degrees of freedom
Multiple R-squared:  0.6035,  Adjusted R-squared:  0.6028
F-statistic:   895 on 2 and 1176 DF, p-value: < 2.2e-16

> anova(grb.mlr)
Analysis of Variance Table

Response: t90
            Df Sum Sq Mean Sq F value    Pr(>F)
```

```
fl                1 497.36   497.36 1382.99 < 2.2e-16 ***
hr32              1 146.36   146.36  406.09 < 2.2e-16 ***
Residuals      1176 422.92     0.36
---
Signif. codes:  0 '***' 0.001 '**' 0.01 '*' 0.05 '.' 0.1 ' ' 1
```

Figure 5.2 displays the scatterplot matrix where we see scatterplots for each pair of variables. The scatterplot in the upper right corner supports a strong linear relationship between *fl* and *t90*—verified by the 0.683 value found in the correlation table. The graph also makes apparent the negative correlation between burst length and burst hardness.

The main diagonal of the matrix shows a density plot for each variable. We can print density plots for individual variables using the *densityPlot* function. Figure 5.3 illustrates the result of applying the densityPlot function to *t90*. We see that most of the bursts have a log length ranging between 1 and 2 seconds.

The *summary* function tells us all three coefficients are significantly different than zero. The R-squared value of 0.6035 together with a significant *F* statistic shows that our model measures around 60% of the variation in the response variable. Lastly, we use the *anova* function to better understand the relevance of each input attribute. Simply, the larger the associated *F* value, the more the variable contributes to the reduction of model error. You can see that the *F* values indicate both variables significantly reduce model error with *fluence* contributing the most to error reduction.

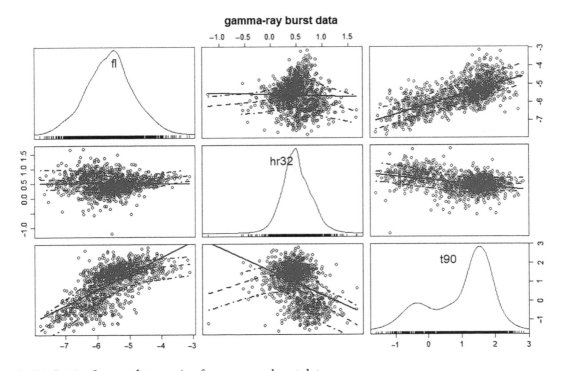

gamma-ray burst data

FIGURE 5.2   Scatterplot matrix of gamma-ray burst data.

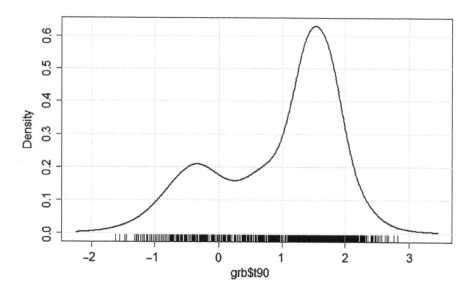

FIGURE 5.3   A density plot of t90.

Once again, all of our analyses are based on the data used to build the model. We have no information about how the regression equation will perform with unseen data. It's time to put our regression model to a real test! First, here are three additional quantitative measures useful for evaluating model performance.

5.2.2 Evaluating Numeric Output

We have seen several useful statistics for evaluating how well a linear regression equation fits the data. Three additional statistics for evaluating models having numeric output are the *mean squared error*, the *root mean squared error*, and the *mean absolute error*. These measures are particularly useful for evaluating how well a model performs when presented with test data.

The *mean squared error* (*mse*) is the average squared difference between actual and computed output, as shown in Equation 5.6.

$$mse = \frac{(a_1 - c_1)^2 + (a_2 - c_2)^2 + \cdots + (a_i - c_i)^2 + \cdots + (a_n - c_n)^2}{n} \tag{5.6}$$

where for the *i*th instance,
   a_i = actual output value
   c_i = computed output value

The *root mean squared error* (*rms*) is simply the square root of the mean squared error. By applying the square root, we reduce the dimensionality of the *mse* to that of the actual error computation. For values less than 1.0, *rms* > *mse*.

You may have noticed that the numerator term in Equation 5.6 is the *residual sum of squares* value used to compute the *residual standard error* (*rse*). The difference between the computation of *rms* and *rse* is in the divisor. With *rms*, the division is by the total number of instances, whereas *rse* is computed by dividing by the degrees of freedom. *rse* is the more rigorous statistic as *rse* > *rms*.

The mean absolute error (*mae*) finds the average absolute difference between actual and computed output values. An advantage of *mae* is that it is less affected by large deviations between actual and computed output. In addition, *mae* maintains the dimensionality of the error value. Equation 5.7 formalizes the definition.

$$mae = \frac{|a_1 - c_1| + |a_2 - c_2| + \cdots + |a_n - c_n|}{n} \tag{5.7}$$

Let's examine a script that uses these measures in a training/test set scenario to evaluate a multiple linear regression model designed to determine burst length.

5.2.3 Training/Test Set Evaluation

Script 5.3 shows the procedure which begins by randomizing and splitting—2/3 for training, 1/3 for testing—the gamma-ray burst data. Next, *lm* is called to create the regression model. The summary function gives the *residual standard error* as 0.6185, a multiple R-squared value of 0.5661, and an adjusted R-squared value of 0.5665. The *F* statistic insures us that the model is not the result of a mere random event.

Next, you see a call to the *mmr.stats* function. This function was written especially for the examples in your text and is found in the *functions.zip* file within your supplementary materials. To make *mmr.stats* part of your RStudio global environment, open *mmr.stats* in the source editor and click *run*. Here's how *mmr.stats* is used.

The parameters given to *mmr.stats* are the regression model, the test file, and the name of the output variable. The function prints values for *mae*, *mse*, *rms*, and *rse* and then returns a four-column table showing actual and predicted values together with the absolute and squared error for each test set instance. Here is the function call as it appears in Script 5.3.

```
output <- mmr.stats(grb.mlr,grb.test,grb.test$t90)
```

mmr.stats uses the base function *predict* which is of particular interest as it is employed multiple times throughout the text. When given a model and a test dataset, *predict* applies the model to the test data and returns a vector containing predictions for the output variable. Predicted output can be numeric or categorical. Here is the call to *predict* where the generic parameters *model* and *test* within *mmr.stats* are replaced with their actual values.

```
predicted <- predict(grb.mlr,grb.test)
```

The statements within *mmr.stats* show how *cbind* and *predict* are used together to create the table of predicted and actual output displayed in Script 5.3.

Returning to Script 5.3, we see the error measures shown in Script 5.3 are not encouraging. However, before answering the question as to whether a linear regression model is acceptable for determining burst length, let's investigate a second method commonly used for evaluating supervised models. We will then use the results of both evaluations to make a decision about how accurately *t90* is predicted by burst brightness and burst hardness.

Script 5.3 Evaluation: Training/Test Set Evaluation

```
> # PREPROCESSING

> # Randomize and split the data: 2/3 training, 1/3 testing
> set.seed(100)
> grb.data <- Grb4u[c(3,4,7)]
> # summary(grb.data)
> index <- sample(1:nrow(grb.data), 2/3*nrow(grb.data))
> grb.train <- grb.data[index,]
> grb.test <-  grb.data[-index,]

> # CREATE THE REGRESSION MODEL
> grb.mlr <- lm(t90 ~ ., data = grb.train)

> #Analyze the results
> summary(grb.mlr)

Call:
lm(formula = t90 ~ ., data = grb.train)

Residuals:
    Min      1Q  Median      3Q     Max
-4.1770 -0.3926  0.0050  0.4057  2.1570

Coefficients:
            Estimate Std. Error t value Pr(>|t|)
(Intercept)  6.05299    0.16795   36.04   <2e-16 ***
fl           0.80970    0.02909   27.84   <2e-16 ***
hr32        -1.15527    0.07492  -15.42   <2e-16 ***
---
Signif. codes:  0 '***' 0.001 '**' 0.01 '*' 0.05 '.' 0.1 ' ' 1

Residual standard error: 0.6185 on 783 degrees of freedom
Multiple R-squared:  0.5661,  Adjusted R-squared:  0.565
F-statistic: 510.8 on 2 and 783 DF,  p-value: < 2.2e-16

> # TEST THE MODEL
> output <- mmr.stats(grb.mlr,grb.test,grb.test$t90)

[1] "Mean Absolute Error ="
[1] 0.4408
[1] "Mean Squared Error ="
```

```
[1] 0.3156
[1] "Root Mean Squared Error="
[1] 0.5618
[1] "Residual standard error="
[1] 0.5646

> head(output)
   actual predicted abs.error sqr.error
3  1.3167    1.0983    0.2184    0.0477
5  1.4879    0.8288    0.6591    0.4344
7  1.2343    0.9229    0.3114    0.0970
8  1.5528    1.0311    0.5217    0.2722
15 1.2076    1.0822    0.1253    0.0157
20 1.3661    1.1471    0.2190    0.0480
```

5.2.4 Using Cross Validation

Cross validation is often the preferred method for evaluating the generalizability of a regression equation. This is especially true when ample test data is not available. Script 5.4 uses our now familiar gamma-ray burst dataset to illustrate one method that applies two functions from the *caret*—classification and regression training—*package* for performing cross validation. Let's take a look at the process. First, be sure to download the *caret package* from the CRAN library.

Script 5.4 starts by dividing the data for training and testing. The training data is employed for the cross validation. The *trainControl* function controls the type of training to take place. For a basic tenfold cross validation, we set method to *cv* and *number* to 10. The *train* function then executes and controls the cross validation by repeatedly calling *lm* using the procedure specified by *trainControl*.

When training is complete, the summary gives a partial list of the ten sample sizes used for the cross validation. There are 1179 total instances in the dataset. We designated 2/3 for training giving a training set of 786 instances. As 10 does not divide 786 evenly, we see variations in the number of instances within the 9 folds used for model building. RMSE and MAE represent average error computed over the 10 folds.

Lastly, the best model built with all training data is then given the test data. The script lists the actual and computed output for the first few instances. The test set *mae* of 0.4408 and *rms* of 0.5618 corroborate the results seen with the training/test set outcome. With a mean absolute error of over four tenths of a second, we cannot be assured of an accurate determination of burst length. To conclude, a nonlinear approach should be considered for our next attempt at determining gamma-ray burst length using burst brightness and hardness measures.

Script 5.4 Evaluation: Cross Validation

```
> library(caret)
Loading required package: lattice
```

```
Loading required package: ggplot2> # PREPROCESSING

> # PREPROCESSING
> # Randomize and split the data for 2/3 training, 1/3 testing
> set.seed(100)
> grb.data <- Grb4u[c(3,4,7)]
> index <- sample(1:nrow(grb.data), 2/3*nrow(grb.data))
> grb.train <- grb.data[index,]
> grb.test <-  grb.data[-index,]

> #PERFORM 10-FOLD CROSS VALIDATION

> xval.control <- trainControl(method = "cv",number = 10)

> # Perform a 10-fold cross validation and return the best model
> # built with all of the training data.

> lm.xval <- train(t90 ~ ., data = grb.train, method =
+                   "lm", trControl = xval.control)
> lm.xval
Linear Regression

No pre-processing
Resampling: Cross-Validated (10 fold)
Summary of sample sizes: 707, 707, 709, 707, 708, 707, ...
Resampling results:

  RMSE        Rsquared    MAE
  0.6177994   0.5700584   0.4767234


> #TEST THE MODEL
> output <- mmr.stats(lm.xval,grb.test,grb.test$t90)
[1] "Mean Absolute Error ="
[1] 0.4408
[1] "Root Mean Squared Error="
[1] 0.5618

> head(output,5)
   actual predicted abs.error sqr.error
3  1.3167   1.0983    0.2184    0.0477
5  1.4879   0.8288    0.6591    0.4344
7  1.2343   0.9229    0.3114    0.0970
8  1.5528   1.0311    0.5217    0.2722
15 1.2076   1.0822    0.1253    0.0157
```

5.2.5 Linear Regression with Categorical Data

Linear regression can also be used to build models for datasets with input variables having categorical values. To see this, before turning your attention to logistic regression, let's apply multiple linear regression to a small subset of the *Ouch's Back & Fracture Clinic (OFC)* dataset used for the case study in Chapter 12. The data is real, but the clinic name is fictitious! This complete dataset poses several challenges as it contains 96 attributes, supports both categorical and numeric data, and contains as wealth of missing data items. Our aim here is for you to become familiar with the dataset and to show you methods for updating and comparing two or more linear models.

5.2.5.1 Ouch's Back and Fracture Clinic Data

Arthur Jones, developer of the Nautilus Exercise Machines, founded the MedX Corporation (http://medxonline.net/) in 1972 with the sole purpose of designing machines to test and strengthen the muscles of the knee and lower back. His theory claims a measureable inverse relationship between pain and strength as well as pain and flexibility. Figure 5.4 shows the pelvic restraint system employed by the MedX lumbar extension machine designed to isolate and strengthen the lumbar spine.

The case study in Chapter 12 experiments with a dataset consisting of 1330 patient records (737 female) who underwent aggressive physical therapy using the MedX machines for treatment of lower back injuries. One of several goals of the case study was to seek relationships identifying patients who prior to treatment are not likely to successfully complete their treatment program. Here we conduct a small pilot study with a subset of 48 male patients together with 12 of the 96 attributes taken from the original data. Our goal is to determine the feasibility of using multiple linear regression with this subset of data to predict expected end of treatment pain levels prior to the start of treatment. In this way,

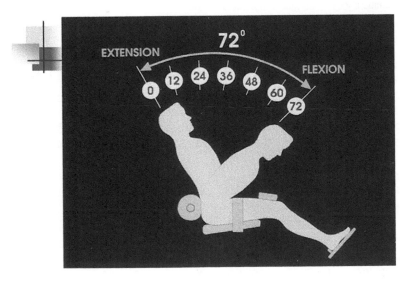

FIGURE 5.4   Pelvic restraint helps isolate and strengthen the lumbar spine.

both patient and therapist will have some initial expectation about treatment success. To begin, import the patient data found in OFCM.csv into RStudio.

Script 5.5 displays the steps along with edited output for this experiment. The first action is attribute removal. As all data instances are male, the gender attribute is useless. Also, *visits*, *LE.rom.out*, and *LE.wt.out* represent end of treatment values and must be removed. *summary(ofc.data)* provides descriptive statistics for the remaining seven input attributes along with the output attribute—*oswest.out*.

Attributes *age*, *height*, and *weight* are self-explanatory. *INS* is a code representing the type of patient health insurance. The original data incorrectly designates *INS* as integer. The lumbar extension weight in (*LE.wt.in*) and lumbar extension range of motion in (*LE.rom.in*) attributes provide lower back strength and flexibility measurements for each patient prior to treatment. These are computerized values taken while the patient initially uses the MedX low back machine. *Oswest.in* and *oswest.out* are positive integers denoting pain level prior to (*oswest.in*) and after (*oswest.out*) completing the treatment program. These scores are somewhat subjective as they are obtained from a questionnaire completed by the patient. An *oswest.out* value less than 20 is generally considered a successful outcome.

The next order of business is to examine attribute correlations (not displayed in the script). Correlations of interest include a correlation of 0.646 between *oswest.in* and *oswest.out*, a correlation of -0.448 between *LE.wt.in* and *oswest.in*, and a correlation of 0.446 between *LE.wt.in* and *LE.rom.in*. The negative correlation between *LE.wt.in* and *oswest.in* supports an inverse relationship between lower back strength and pain level.

Script 5.5 Modeling Treatment Outcome

```
> #PREPROCESSING
> # Remove gender & end of treatment attributes
> ofc.data <- OFCM[-c(1,6,8,10)]
> #Data Summary
> summary(ofc.data)

      age             height          weight             INS           LE.rom.in
 Min.   :18.00   Min.   :40.00   Min.   :152.0   Min.   :1.000   Min.   :24.00
 1st Qu.:33.75   1st Qu.:69.00   1st Qu.:183.2   1st Qu.:1.000   1st Qu.:35.5
 Median :43.00   Median :71.00   Median :203.0   Median :2.000   Median :45.0
 Mean   :43.56   Mean   :70.31   Mean   :211.2   Mean   :2.375   Mean   :43.6
 3rd Qu.:52.00   3rd Qu.:72.00   3rd Qu.:226.2   3rd Qu.:4.000   3rd Qu.:51.5
 Max.   :76.00   Max.   :77.00   Max.   :322.0   Max.   :5.000   Max.   :72.0

    LE.wt.in         oswest.in        oswest.out
 Min.   : 40.00   Min.   : 0.00   Min.   : 0.00
 1st Qu.: 63.75   1st Qu.:23.50   1st Qu.: 6.00
 Median : 75.00   Median :32.00   Median :16.00
 Mean   : 84.92   Mean   :31.92   Mean   :17.58
 3rd Qu.:120.00   3rd Qu.:40.00   3rd Qu.:24.00
 Max.   :172.00   Max.   :60.00   Max.   :54.00

> # Check for correlations
```

```
> #round(cor(ofc.data),3)

> set.seed(100)
> # convert insurance to character
> ofc.data$INS <- as.character(ofc.data$INS)
>
> CREATE REGRESSION MODEL
> osw.mlr <- lm(oswest.out ~ ., data =ofc.data, na.action=na.omit)
> summary(osw.mlr)

Call:
lm(formula = oswest.out ~ ., data = ofc.data,)

Residuals:
    Min      1Q  Median      3Q     Max
-22.921  -6.691  -2.314   6.476  23.249

Coefficients:
              Estimate Std. Error t value Pr(>|t|)
(Intercept) -70.91163   39.21754  -1.808   0.0787 .
age           0.01803    0.16938   0.106   0.9158
height        0.96447    0.54596   1.767   0.0855 .
weight       -0.02305    0.06115  -0.377   0.7084
INS2         11.98744   13.56683   0.884   0.3826
INS3         -1.65482    6.40556  -0.258   0.7976
INS4          4.56606    4.65065   0.982   0.3326
INS5          1.80947   10.44490   0.173   0.8634
LE.rom.in     0.07858    0.22419   0.351   0.7279
LE.wt.in     -0.05389    0.08343  -0.646   0.5223
oswest.in     0.75981    0.15831   4.799 2.62e-05 ***
---
Signif. codes:  0 '***' 0.001 '**' 0.01 '*' 0.05 '.' 0.1 ' ' 1

Residual standard error: 12.14 on 37 degrees of freedom
Multiple R-squared:  0.4866,  Adjusted R-squared:  0.3478
F-statistic: 3.507 on 10 and 37 DF, p-value: 0.002492

> head(mmr.stats(osw.mlr,ofc.data,ofc.data$oswest.out))
[1] "Mean Absolute Error ="
[1] 8.4442
[1] "Mean Squared Error ="
[1] 113.6326
[1] "Root Mean Squared Error="
[1] 10.6599
[1] "Residual standard error="
[1] 12.1415
  actual predicted abs.error sqr.error
```

1	54	33.7426	20.2574	410.3606
2	54	37.6762	16.3238	266.4670
3	50	32.1052	17.8948	320.2243
4	48	41.5605	6.4395	41.4671
5	42	27.2523	14.7477	217.4933
6	40	16.7513	23.2487	540.5012

The statement following *set.seed* changes the data type of *INS* to character. R handles the conversion of a variable having *v* values to character type by creating *v*-1 auxiliary variables. For each data instance, the auxiliary variables will be either 0 or 1. For our example, four new variables—*INS2, INS3, INS4,* and *INS5*—are created. If the original value of *INS* = 2, the new variable named *INS2* will have value 1 with *INS3, INS4, INS5* all showing 0. The same reasoning applies to *INS3, INS4,* and *INS5*. If the original value shows *INS* = 1, all of the auxiliary variables will be 0.

Next, the call to *lm* includes *na.action=na.omit*. The value *na.omit* tells us that any instance with at least one missing value will be omitted. As *na.omit* is the default setting for *na.action*, the use of this function is only needed when some action other than omitting instances with missing values is desired.

The summary function shows that *oswest.in* is the only regression coefficient significantly different from 0. Also, a *residual standard error* of 12.14 for an outcome variable ranging between 0 and 54 is less than optimal. However, the *p-value* of 0.0025 associated with the *F statistic* indicates that at least one input variable determines the output beyond chance. There are at least three options for obtaining a better result:

- Use the *update* function to make minor changes to the current model.

- Use the *step* function to remove attributes one at a time in order to find an "optimal" solution.

- Look for outliers in the data that may be negatively influencing the results. Delete the outliers and build a new model.

Let's examine each option.

5.2.5.2 The Update Function

The edited output given in Script 5.6 displays one procedure for applying the *update* function. First, we use the *anova* function to obtain informative statistics about each input variable. Of primary importance is the effect each variable has on the reduction of the sum of squared errors—the Sum Sq column. Those attributes contributing the least are candidates for removal. The attribute of least importance in this regard is *height* with *weight* a close second. Notice that *oswest.in* and *LE.wt.in* are the only attributes shown to be significant.

Once the *update* function is applied, we use the *anova* function in a different capacity. Specifically, we use the function to compare the original and new models. The comparison shows the new model reduces the sum of square error by 460.05. Unfortunately, the F statistic of 0.0856 computed for the reduction is not significant.

Script 5.6 The Update Function

```
> # EXAMINE ATTRIBUTE SIGNIFICANCE
> anova(osw.mlr)
Analysis of Variance Table
```

Response: oswest.out

	Df	Sum Sq	Mean Sq	F value	Pr(>F)	
age	1	35.0	35.0	0.2373	0.62905	
height	1	4.4	4.4	0.0299	0.86363	
weight	1	14.2	14.2	0.0965	0.75784	
INS	4	520.4	130.1	0.8826	0.48383	
LE.rom.in	1	221.5	221.5	1.5023	0.22807	
LE.wt.in	1	978.2	978.2	6.6355	0.01412	*
oswest.in	1	3395.7	3395.7	23.0346	2.617e-05	***
Residuals	37	5454.4	147.4			

```
---
Signif. codes:  0 '***' 0.001 '**' 0.01 '*' 0.05 '.' 0.1 ' ' 1
```

```
> # REMOVE LEAST SIGNIFICANT ATTRIBUTE
> osw2.mlr <- update(osw.mlr, .~ . -height)
> # summary(osw2.mlr)
```

```
> # COMPARE THE MODELS
> anova(osw.mlr,osw2.mlr)
Analysis of Variance Table
```

```
Model 1: oswest.out ~ age + height + weight + INS + LE.rom.in +
LE.wt.in + oswest.in
Model 2: oswest.out ~ age + weight + INS + LE.rom.in + LE.wt.in +
oswest.in
```

	Res.Df	RSS	Df	Sum of Sq	F	Pr(>F)	
1	37	5454.4					
2	38	5914.4	-1	-460.05	3.1208	0.08555	.

```
---
Signif. codes:  0 '***' 0.001 '**' 0.01 '*' 0.05 '.' 0.1 ' ' 1
```

5.2.5.3 The Step Function

We could continue with the *update* function removing one feature at a time but a better approach is to try the *step* function. By default, the *step* function takes our original model and performs a backward elimination to create a best model. The best model is defined as the one with the minimum value for the Akaike Information Criterion (AIC). *AIC* computes its score for each model by weighting goodness of model fit against model complexity.

Script 5.7 shows the call to the *step* function together with the result of comparing the original model with the best model. Script 5.75 (not displayed) compares the original model with a model that combines decimal scaling with the *step* function. Both comparisons show no significant model improvement.

Script 5.7 The Step Function

```
> # Use the step function to create a final model
> oswFinal.mlr <- step(osw.mlr)
> summary(oswFinal.mlr)
> anova(oswFinal.mlr)
> anova(osw.mlr,oswFinal.mlr)


Model 1: oswest.out ~ age + height + weight + INS + LE.rom.in +
LE.wt.in + oswest.in
Model 2: oswest.out ~ height + oswest.in
  Res.Df    RSS Df Sum of Sq      F Pr(>F)
1     37 5454.4
2     45 5848.7 -8   -394.33 0.3344  0.947
```

5.2.5.4 Checking for Outliers

Let's try our third option with a quick check for outliers. We can graphically depict candidate outliers with the *boxplot* function. Namely,

```
>boxplot(ofc.data$oswest.out,horizontal = T,main="Oswestry.out")
```

Figure 5.5 shows the boxplot defined by a minimum, lower quartile (25%), median, upper quartile (75%), and a maximum. These values are respectively 0, 6, 16, 24, and 54. Outliers lie outside 1.5 times above or below the interquartile range (24 – 16). For our example, two instances with an *oswestry.out* score above 51 qualify as outliers. We can also zero in on candidate outliers by applying the *car* package function *outlierTest* to our original model. *outlierTest* is useful but is also a lot of work as each call gives us the best candidate outlier. Here is the test.

```
>library(car)
>outlierTest(osw.mlr)
Studentized residuals with Bonferroni p < 0.05
Largest |rstudent|:
   rstudent unadjusted p-value Bonferroni p
6   2.236216           0.031627           NA
```

The output indicates the 6th instance as an outlier. The 6th instance is an individual whose pain level increased from 32 to 40 upon completion of treatment. Certainly an outlier! Repeating this process (not shown) finds instances 1, 41, 3 and several others as candidate outliers. Once again, removing the outliers and rebuilding the regression model does not lead to a significantly better result.

Oswestry.out

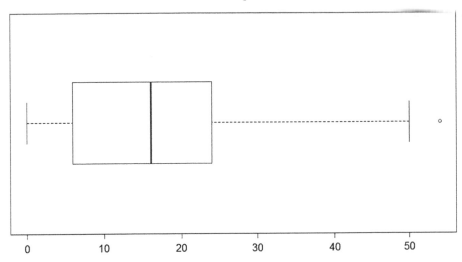

FIGURE 5.5   Boxplot of oswestry.out.

Finally, we must be skeptical of any conclusions obtained from a dataset of this size. We must also keep in mind that this analysis is based only on training data. However, the examples and analyses given here have provided you with several tools to use at the next stop on your tour of machine learning with R. It's time to move on to logistic regression!

5.3 LOGISTIC REGRESSION

In the previous section, you saw how linear regression can be applied to datasets having input variables of numeric and categorical type. Linear regression has also been used when the dependent (output) variable represents a binary outcome. To illustrate, suppose the dependent variable is *diagnosis* having values *healthy* or *sick*. In order to use linear regression, we must have a numeric output so we replace healthy with *0* and sick with *1*. However, the general methodology of using linear regression to model problems with observed outcome restricted to two values is seriously flawed.

The problem lies in the fact that the value restriction placed on the dependent variable is not observed by the regression equation. That is because linear regression produces a straight-line function; values of the dependent variable are unbounded in both the positive and negative directions. Therefore, for the right-hand side of Equation 5.1 to be consistent with a binary outcome, we must transform the linear regression model. By transforming the linear model so as to restrict values for the output attribute to the [0, 1] interval range, the regression equation can be thought of as producing a probability of the occurrence or nonoccurrence of a measured event.

Although several options for transforming the linear regression model exist, we restrict our discussion to the logistic model. The logistic model applies a logarithmic transformation that makes the right-hand side of Equation 5.1 an exponential term in the transformed equation.

5.3.1 Transforming the Linear Regression Model

Logistic regression is a nonlinear regression technique that associates a conditional probability score with each data instance. To understand the transformation performed by the logistic model, we begin by thinking of Equation 5.1 as computing a probability. A probability value of 1 denotes the observation of one class (e.g., *diagnosis = healthy*). Likewise, a probability of 0 indicates observance of the second class (e.g., *diagnosis = sick*). Equation 5.8 is a modified form of Equation 5.1 where the left-hand side of the equation is written as a conditional probability.

$$p(y=1|\mathbf{x}) = a_1 x_1 + a_2 x_2 + a_3 x_3 + \cdots + a_n x_n + c \tag{5.8}$$

Equation 5.8 shows $p(y = 1 | \mathbf{x})$ as an unbounded value denoting the conditional probability of seeing the class associated with $y = 1$ given the values contained in feature (attribute) vector $\mathbf{x}$. To eliminate the boundary problem seen in the equation, the probability is transformed into an odds ratio. Specifically,

$$\left(\frac{p(y=1|\mathbf{x})}{1-p(y=1|\mathbf{x})} \right) \tag{5.9}$$

For any feature vector $\mathbf{x}$, the odds indicate how often the class associated with $y = 1$ is seen relative to the frequency in which the class associated with $y = 0$ is observed. The natural log of this odds ratio (known as the *logit*) is then assigned to the right-hand side of Equation 5.8 That is,

$$\ln\left(\frac{p(y=1|\mathbf{x})}{1-p(y=1|\mathbf{x})} \right) = \mathbf{a}\mathbf{x} + c \tag{5.10}$$

where
$$\mathbf{x} = (x_1, x_2, x_3, x_4, \ldots, x_n);$$
$$\mathbf{a}\mathbf{x} + c = a_1 x_1 + a_2 x_2 + a_3 x_3 + \ldots + a_n x_n + c$$

Finally, we solve Equation 5.10 for $p(y = 1 | \mathbf{x})$ to obtain a bounded representation for the probability, which is shown in Equation 5.11.

$$p(y=1|x) = \frac{e^{ax+c}}{1+e^{ax+c}} \tag{5.11}$$

where
  e is the base of natural logarithms often denoted as exp.

5.3.2 The Logistic Regression Model

Equation 5.11 defines the logistic regression model. Figure 5.6 shows that the graph of the equation is an s-shaped curve bounded by the [0, 1] interval range. As the exponent term

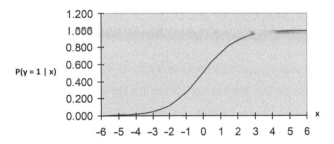

FIGURE 5.6 The logistic regression equation.

approaches negative infinity, the right-hand side of Equation 5.11 approaches 0. Likewise, as the exponent becomes infinitely large in the positive direction, the right side of the equation approaches 1.

The method used to determine the coefficient values for the exponent term **ax** + c in Equation 5.11 is iterative. The purpose of the method is to minimize the sum of logarithms of predicted probabilities. Convergence occurs when the logarithmic summation is close to 0 or when the value does not change from one iteration to the next. Details of the technique are beyond the scope of this book. For the interested reader, a description of the process is available from several sources (Hosmer and Lemeshow, 1989; Long, 1989).

5.3.3 Logistic Regression with R

Now that you've seen the mathematics, let's apply logistic regression to the mixed form of the cardiology patient dataset. Recall that this dataset contains information about 303 individuals 138 of which have had a heart attack. You can use the *str* function to verify that five of the thirteen input attributes are factor variables, two are logical, and five are of integer type. The output variable is a factor of two levels: *healthy* (1) or *sick* (2).

Script 5.8 shows the process and partial output for our experiment. Preprocessing sets the stage for a training/test set scenario. Next is the call to *glm* (generalized linear model) which can be used to implement several models including logistic regression. The format is similar to *lm* but with added parameters for specifying model choice. As logistic regression requires a binary output variable, 0 replaces the level associated with 1 (*healthy*) and 1 replaces the level associated with 2 (*sick*).

The summary function gives the coefficients for the created model. Several nonsignificant coefficients have been removed to conserve space. Notice that *number of color vessels* provides the most significant contribution to the regression equation.

With generalized linear models, the *chi-square* version of the *anova* function is used for determining if the addition of a new variable significantly reduces model deviance. Deviance reduction in exponential models (ex. logistic regression) is similar to reducing the sum of squares in linear models. Script 5.8 gives only those attributes that significantly reduce the deviance—you can examine the output of your script to get an accurate measure of individual reductions in deviance. Notice that there is some disagreement between the variables determined to have coefficients significantly different from zero and the variables that significantly reduce model deviance levels.

The *predict* function applies the logistic model to the test data. Type = 'response' gives the output as predicted probabilities. For each test set instance, *card.results* will be a numeric value between 0 and 1 inclusive. To see this, *cbind* is used to create a table showing predicted values and integers representing factor levels. The *head* function lists the first six table instances. The first test set instance is the sixth instance within the original data. The predicted output value of 0.003 is correct as the actual classification is *healthy* (1). Namely, *card.results* values greater than 0.50 classify with the *sick* class, whereas *card.results* values less than or equal to 0.50 classify with the *healthy* class. Clearly, we can be more confident of an association with the *sick* class given a value for *card.results* close to 1.0 as compared to a value of 0.60!

Script 5.8 Logistic Regression with Cardiology Patient Data

```
> #PREPROCESSING

> set.seed(100)
> card.data <- CardiologyMixed
> index <- sample(1:nrow(card.data), 2/3*nrow(card.data))
> card.train <- card.data[index,]
> card.test <-  card.data[-index,]
> c.glm <- glm(class ~ .,data = c.train,family= binomial(link='logit'))
> summary(card.glm)

> # CREATE AND ANALYZE LOGISTIC REGRESSION MODEL

Call:
glm(formula = class ~ ., family = binomial(link ="logit"),data = c.train)

Deviance Residuals:
    Min       1Q     Median       3Q       Max
-2.7797  -0.4401   -0.1139    0.2882    3.0655

Coefficients:
                                 Estimate Std. Error z value
Pr(>|z|)
(Intercept)                      9.050e+00  1.213e+03   0.007 0.994
genderMale                       1.715e+00  7.601e-01   2.256 0.024 *
chest.pain.typeAbnormal Angina  -9.783e-01  7.455e-01  -1.312 0.189
chest.pain.typeAngina           -2.708e+00  9.304e-01  -2.910 0.003 **
chest.pain.typeNoTang           -1.980e+00  6.730e-01  -2.942 0.003 **
maximum.heart.rate              -5.334e-03  1.555e-02  -0.343 0.731
slopeFlat                       -1.488e-01  1.161e+00  -0.128 0.898
slopeUp                         -1.590e+00  1.202e+00  -1.323 0.185
X.colored.vessels                1.306e+00  3.797e-01   3.440 0.000 ***
thalNormal                      -6.654e-01  1.124e+00  -0.592 0.554
thalRev                          1.192e+00  1.082e+00   1.102 0.270
---
```

```
Signif. codes:  0 '***' 0.001 '**' 0.01 '*' 0.05 '.' 0.1 ' ' 1

(Dispersion parameter for binomial family taken to be 1)

    Null deviance: 279.06  on 201  degrees of freedom
Residual deviance: 116.85  on 183  degrees of freedom
AIC: 154.85

Number of Fisher Scoring iterations: 15

> anova(card.glm, test="Chisq")
Analysis of Deviance Table

Response: class

                 Df Deviance Resid. Df Resid. Dev  Pr(>Chi)
NULL                                 201     279.06
age               1   11.864       200     267.20 0.000 ***
gender            1   15.666       199     251.53 7.55e-05***
chest.pain.type   3   58.595       196     192.94 1.17e-12***
maximum.heart.rate 1  11.103       190     172.24 0.000 ***
peak              1   17.909       188     152.84 2.3e-05 ***
X.colored.vessels 1   18.830       185     129.79 1.4e-05 ***
thal              2   12.934       183     116.85 0.001 **
---
Signif. codes:  0 '***' 0.001 '**' 0.01 '*' 0.05 '.' 0.1 ' ' 1

> card.results <- predict(card.glm, card.test, type='response')
> card.table <- cbind(pred=round(card.results,3),Class= card.test$class)
> card.table <- data.frame(card.table)

> head(card.table)
    Pred Class
6  0.003   1
9  0.199   1
10 0.998   2
17 0.568   1
18 0.125   1
21 0.017   1


> # CREATING A CONFUSION MATRIX

> # healthy <= 0.5 sick > 0.5
> card.results <- ifelse(card.results > 0.5,2,1) #  > .5 a sick

> card.pred <- factor(card.results,labels=c("Healthy","Sick"))
> my.conf <- table(card.test$class,card.pred,dnn=c("Actual","Predicted"))
> my.conf
```

```
            Predicted
Actual      Healthy Sick
    Healthy      51     6
    Sick         11    33
```

```
> confusionP(my.conf)
```

```
    Correct= 84  Incorrect= 17  Accuracy = 83.17 %
```

5.3.4 Creating a Confusion Matrix

Chapter 1 showed you how a confusion matrix can be used to evaluate the accuracy of a supervised learner model. It's easy to modify the numeric output of a logistic regression model to create a confusion matrix. Script 5.8 shows how it is done!

To create the confusion matrix, we first convert the computed test set output from a list of probabilities to a list of 1's (*healthy*) and 2's (*sick*). This is accomplished in two steps. First, the *ifelse* statement replaces a probability greater than 0.5 with a 2. Probability values less than 0.05 are replaced with 1. Once the values in *card.results* have been converted to 1's and 2's, we use the *factor* function to replace the 1's and 2's with their associated level (*healthy* or *sick*). The *table* function then creates the confusion matrix—*dnn* specifies the dimension names for the table. The *confusionP* function of Chapter 3 displays a test set accuracy of 83.17%. We see that 6 of the 17 misclassifications are healthy individuals classified as sick. The remaining misclassifications are sick individuals classified as healthy.

In many situations, classification accuracy actually takes a back seat to the distribution of misclassifications. For example, if our goal is to predict customer churn (customers likely to drop their service), a model that shows a high level of general accuracy that is unable to identify churning customers is without value. The next section offers a graphical approach for analyzing the distribution of misclassifications among alternative models (Figure 5.7).

5.3.5 Receiver Operating Characteristics (ROC) Curves

Before looking at receiver operating characteristics (ROC) curves, some confusion matrix terminology is in order. Table 5.1 shows a confusion matrix for a two-class problem where *true positives* and *true negatives* represent correct classifications. *False negatives* depict positive instances that have been incorrectly identified with the negative (*no*) class. Similarly, *false positives* are negative instances that have been incorrectly identified with the positive (*yes*) class. Having these definitions, we can now begin our discussion of *ROC curves*. *ROC curves* are a two-dimensional graphical approach for depicting the tradeoff between true positive rate and false positive rate. ROC curves were first used during World War II for analyzing radar images. Today, ROC graphs are an especially useful visualization and analysis tool in two-class domains having imbalanced or cost-sensitive data.

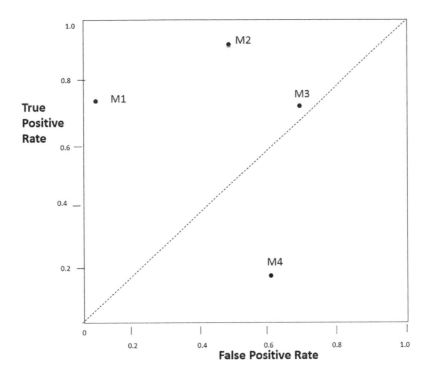

FIGURE 5.7   An ROC graph for four competing models.

TABLE 5.1   Possible Outcomes of a Two-Class Prediction

		Predicted Class	
		Yes	No
Actual class	Yes	True positive (TP)	False negative (FN)
	No	False positive (FP)	True negative (TN)

With ROC curves, the true positive rate is plotted on the y-axis and the false positive rate is plotted on the x-axis. For class C, the *true positive rate* (TP rate) is computed by dividing the number of instances correctly classified with C by the total number of instances actually belonging to C. Specifically, TP rate = TP/(TP + FN). The *false positive rate* (FP rate) for class C is the total number of instances wrongly classified with C divided by the total number of instances not in C. That is, FP rate = FP/(FP + TN). Values for TP rate and FP rate are of special interest when the cost of incorrectly classifying a nonclass instance as a class member is significantly different from the cost of incorrectly classifying a class instance as a member of another class.

Each point on an ROC curve represents one model induced by the classifier. With imbalanced data, the rare class is treated as the positive class as it is of most interest. ROC graphs are best illustrated by example.

The simplest case is given in Figure 5.7 which shows an ROC graph for four competing models. Each model is discrete as it outputs a single class label and is associated with exactly one point of the graph (TP rate, FP rate). Any point along the diagonal line represents random guessing. Let's assume each model is one alternative for detecting credit card fraud. The fraudulent transactions are the rare class and, hence, are represented by the true positive axis. The graph shows that M1 captures over 70% of the fraudulent transactions while accepting less than 10% of false positive instances. M2 secures around 90% of the true positives but also accepts approximately 50% of the false positives. M3 performs a little better than random guessing and M4 gives a false positive rate of more than 60%.

ROC curves are not able to tell us which model is the best choice as we are unaware of the costs associated with incorrect classifications. However, the ROC graph can be a key component of the analytics process especially in domains involving rarity. Let's continue with our example by creating an ROC curve for modeling cardiology patient data. Script 5.9 provides the details!

Script 5.9 ROC Curve for Cardiology Patient Data

```
> # CREATE THE ROC CURVE AND DETERMINE THE AUC

> library(ROCR)
> p.card <-predict(card.glm, card.test, type="response")
> pr.card <- prediction(p.card, card.test$class)
> pr.card # Uses slots to create the information needed by
performance
> prf.card <- performance(pr.card, measure="tpr", x.measure="fpr")
> plot(prf.card)

> # DETERMINE THE AREA UNDER THE CURVE
> auc <- performance(pr.card, measure = "auc")
> auc <- auc@y.values[[1]]
> auc
[1] 0.895933
```

To begin, you will need to install the ROCR package found within the CRAN library. Script 5.9 begins by repeating the predictions for the test data. The predictions are then given to *prediction* which among other things uses slots to hold information needed by the *performance* function whose job it is to create the ROC structure. To see this, scroll the output of *pr.card* until you locate slots named *predictions*, *labels*, *cutoffs*, *tp*, and *fp*.

- *Predictions*. Lists the prediction probabilities as determined by the model for each item in the test set.

- *Labels*. Lists the actual class for each test set instance. Upon examining the first 10 instances, we see that instances 4, 9, and 10 are incorrect predictions.

- *Cutoffs*. Lists the prediction probabilities in sorted descending order.

- *Tp & fp*. These slots are used to plot the ROC curve. Horizontal moves represent misclassifications.

Here are edited contents for the first few values in each of the aforementioned slots.

```
Slot "predictions": (original ordering of the test data)

0.003474688 0.199175170 0.997912546 0.568080750 0.125149902
0.017199134 0.002664405 0.040005274 0.026041205 0.805670532

Slot "labels": (original ordering of the test data)

Healthy Healthy Sick Healthy Healthy Healthy Healthy Healthy Sick
Healthy

Slot "cutoffs": (Predictions sorted in descending order)

0.999964675 0.999748982 0.998960310 0.997912546 0.997557427
0.99633896 0.991093656 0.990741795 0.989831370 0.989775985

Slot "fp":

0 1 1 1 1 1 1 1 1 1 1 1 1 1 1 2 2 2…

Slot "tp":

0 0 1 2 3 4 5 6 7 8 9 10 11 12 13 14 14 15 16…
```

You can display the ROC curve by executing the script through the *plot* function. Table 5.2 has been constructed to match the definitions in Table 5.1 and emphasize that the group of individuals who have had at least one heart attack represents the *positive* class. Let's take a look at how the graph is constructed.

Both *tp* and *fp* show *0* as the starting position in the ROC graph. Next, we see a horizontal move as *fp* moves from 0 to 1. This tells us that first instance with the highest probability of being from the sick class is actually an individual who according to the data has not had a heart attack! After this horizontal move, *tp* shows 14 unique values (1–14). This tells us the graph makes 14 vertical moves representing true positive predictions. Notice that the corresponding values for *fp* are all *1* indicating no horizontal movement. Next 14 is repeated and *fp* moves from 1 to 2 giving another false positive prediction. This process continues until the graph terminates at the point where both TP and FP rates are 1. All instances are classified with the positive (sick) class. Figure 5.8 highlights the point on the ROC graph for our model where TP rate = 0.750 (33/44) and FP rate = 0.105 (6/57).

TABLE 5.2 Confusion Matrix from Script 5.8 Designating Sick Individuals as the Positive Class

		Predicted Class	
		Sick	Healthy
Actual class	Sick	33	11
	Healthy	6	51

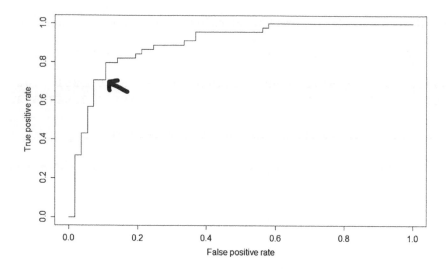

FIGURE 5.8   An ROC curve created by Script 5.9.

At this point, you may be asking how all of this information can be put to good use. Well, let's suppose the price for incorrectly identifying individuals who have had a heart attack as healthy is very high. Table 5.2 shows that 11 test set individuals (false negatives) fall into this category. The ROC graph allows us to clearly see the false positive classification price we must pay in order to move some of the false negatives into the positive column. For example, in order to move the TP rate from 0.75 to 0.80, we will see a small change in the false positive rate. The question as to whether this should be done is one for the researcher.

A second question is how do we modify the model so as to increase the true positive rate? For the answer, we return to Script 5.8 and modify the ifelse statement. For example, changing the statement as follows:

```
card.results <- ifelse(card.results > 0.4,2,1
```

lowers the inclusion criterion for the class *sick* class, thereby increasing the true positive rate. Making this change (not shown) to the script increases the TP rate to 0.795, the FP rate to 0.123 and increases overall classification accuracy to 84.16%.

5.3.6 The Area under an ROC Curve

The area under an ROC curve (*auc*) is also of interest as it provides a quantitative measure for evaluating model performance. To see this, return to Script 5.9 where the *performance* function is now used to compute *auc*. As true positive and false positive values range between 0 and 1, the maximum possible value for *auc* is 1. In this case, the graph immediately moves vertically where all true positive instances are correctly classified without a single false positive error.

The computed *auc* for our example is 0.8595 which tells us that the majority of the area under the curve is covered. This means that the ROC curve tends to move vertically rather than in a horizontal direction, thereby identifying true positive rather than false positive instances. Several end-of-chapter exercises provide additional experiments focusing on ROC curves. Next on our list of supervised statistical models is the Naïve Bayes classifier!

5.4 NAÏVE BAYES CLASSIFIER

The *Naïve Bayes classifier* offers a simple yet powerful supervised classification technique. The classifier is termed naïve as it assumes the independence of all input attributes, and that numeric attribute values are normally distributed. Naïve Bayes usually performs very well even when these assumptions are known to be false. The classifier is based on *Bayes theorem*, which is stated as:

$$P(H|E) = \frac{P(E|H) \times P(H)}{P(E)} \tag{5.12}$$

where *H* is a hypothesis to be tested and *E* is the evidence associated with the hypothesis.

From a classification viewpoint, the hypothesis is the dependent variable and represents the predicted class. The evidence is determined by values of the input attributes. $P(E|H)$ is the *conditional probability* that *E* is true given *H*. $P(H)$ is an *a priori probability*, which denotes the probability of the hypothesis before the presentation of any evidence. Conditional and a priori probabilities are easily computed from the training data. The classifier is best understood with an example.

5.4.1 Bayes Classifier: An Example

Consider the data in Table 5.3, which is a subset of the credit card promotion database defined in Chapter 1. For our example, we use *gender* as the output attribute whose value is to be predicted. Table 5.4 lists the distribution (counts and ratios) of the output attribute values for each input attribute. To demonstrate, Table 5.4 tells us that four males took advantage of the magazine promotion, and that these four males represent two-thirds of the total male population. As a second example, Table 5.4 indicates that three of the four female dataset instances purchased the magazine promotion.

Let's use the data in Table 5.4 together with the Bayes classifier to perform a new classification. Consider the following new instance to be classified:

Magazine Promotion = Yes

Watch Promotion = Yes

Life Insurance Promotion = No

Credit Card Insurance = No

Gender= ?

We have two hypotheses to be tested. One hypothesis states the credit card holder is male. The second hypothesis sees the instance as a female card holder. In order to determine which hypothesis is correct, we apply the Bayes classifier to compute a probability for

TABLE 5.3   Data for the Bayes Classifier

Magazine Promotion	Watch Promotion	Life Insurance Promotion	Credit Card Insurance	Gender
Yes	No	No	No	Male
Yes	Yes	Yes	Yes	Female
No	No	No	No	Male
Yes	Yes	Yes	Yes	Male
Yes	No	Yes	No	Female
No	No	No	No	Female
Yes	Yes	Yes	Yes	Male
No	No	No	No	Male
Yes	No	No	No	Male
Yes	Yes	Yes	No	Female

TABLE 5.4   Counts and Probabilities for Attribute *Gender*

Gender	Magazine Promotion		Watch Promotion		Life Insurance Promotion		Credit Card Insurance	
	Male	Female	Male	Female	Male	Female	Male	Female
Yes	4	3	2	2	2	3	2	1
No	2	1	4	2	4	1	4	3
Ratio: yes/total	4/6	3/4	2/6	2/4	2/6	3/4	2/6	1/4
Ratio: no/total	2/6	1/4	4/6	2/4	4/6	1/4	4/6	3/4

each hypothesis. The general equation for computing the probability that the individual is a male customer is:

$$P(gender = male|E) = \frac{P(E|gender = male) \times P(gender = male)}{P(E)}$$

(5.13)

Let's start with the conditional probability P(E | gender = male). This probability is computed by multiplying the conditional probability values for each piece of evidence. This is possible if we make the assumption that the evidence is independent. The overall conditional probability is the product of the following four conditional probabilities:

P(magazine promotion = yes | gender = male) = 4/6

P(watch promotion = yes | gender = male) = 2/6

P(life insurance promotion = no | gender = male) = 4/6

P(credit card insurance = no | gender = male) = 4/6

These values are easily obtained as they can be read directly from Table 5.2. Therefore, the conditional probability for gender = male is computed as:

$$P(E \mid gender = male) = (4/6)\,(2/6)\,(4/6)\,(4/6)$$

$$= 8/81$$

The a priori probability, denoted in Equation 5.13 as $P(gender = male)$, is the probability of a male customer without knowing the promotional offering history of the instance. In this case, the a priori probability is simply the fraction of the total population that is male. As there are six males and four females, the a priori probability for $gender = male$ is 3/5.

Given these two values, the numerator expression for Equation 5.13 becomes:

$$(8/81)\,(3/5) \cong 0.0593$$

We now have,

$$P(gender = male \mid E) \cong 0.0593/P(E)$$

Next, we compute the value for $P(gender = female \mid E)$ using the formula:

$$P(gender = female \mid E) = \frac{P(E \mid gender = female) \times P(gender = female)}{P(E)} \quad (5.14)$$

We first compute the conditional probability with values obtained directly from Table 5.4. Specifically,

$P(magazine\ promotion = yes \mid gender = female) = 3/4$

$P(watch\ promotion = yes \mid gender = female) = 2/4$

$P(life\ insurance\ promotion = no \mid gender = female) = 1/4$

$P(credit\ card\ insurance = no \mid gender = female) = 3/4$

The overall conditional probability is:

$$P(E \mid gender = female) = (3/4)\,(2/4)\,(1/4)\,(3/4)$$

$$= 9/128$$

As there are four females, the a priori probability for $P(gender = female)$ is 2/5.

Therefore, the numerator expression for Equation 5.14 becomes:

$$(9/128)\,(2/5) \cong 0.0281$$

We now have

$$P(gender = female \mid E) \cong 0.0281/P(E)$$

Finally, we need not concern ourselves with $P(E)$ because it represents both the probability of the evidence when *gender* = *male* and *gender* = *female*. Therefore because 0.0593 > 0.0281, the Bayes classifier tells us the instance is most likely a male credit card customer.

5.4.2 Zero-Valued Attribute Counts

A significant problem with the Bayes technique is when one of the counts for an attribute value is 0. For example, suppose the number of females with a value of *no* for *credit card insurance* is 0. In this case, the numerator expression for $P(gender = female \mid E)$ will be 0. This means that the values for all other attributes are irrelevant because any multiplication by 0 gives an overall 0 probability value.

To solve this problem, we add a small constant, k, to the numerator (n) and denominator (d) of each computed ratio. Therefore, each ratio of the form n/d becomes

$$\frac{n+(k)(p)}{d+k} \tag{5.15}$$

where
  k is a value between 0 and 1 (usually 1)

  p is chosen as an equal fractional part of the total number of possible values for the attribute. (For example, if an attribute has two possible values, p will be 0.5.) An alternative is to simply add one to each attribute value count in order to avoid zero probabilities. This type of modification is often referred to as *Laplace smoothing*.
Let's use the first technique to recompute the conditional probability $P(E \mid gender = female)$ for our previous example. With $k = 1$ and $p = 0.5$, the conditional probability of the evidence given *gender* = *female* computes to:

$$\frac{(3+0.5)\times(2+0.5)\times(1+0.5)\times(3+0.5)}{5\times5\times5\times5} \cong 0.0176$$

5.4.3 Missing Data

Fortunately, missing data items are not a problem for the Bayes classifier. To demonstrate, consider the following instance to be classified by the model defined in Table 5.4.

  Magazine Promotion = Yes

  Watch Promotion = Unknown

  Life Insurance Promotion = No

  Credit Card Insurance = No

  Gender = ?

As the value of *watch promotion* is unknown, we can simply ignore this attribute in our conditional probability computations. By doing so, we have:

$P(E \mid gender = male) = (4/6)\ (4/6)\ (4/6) = 8/27$

$P(E \mid gender = female) = (3/4)\ (1/4)\ (3/4) = 9/64$

$P(gender = male \mid E) \cong 0.1778/P(E)$

$P(gender = female \mid E) \cong 0.05625/P(E)$

As you can see, the effect is to give a probability value of 1.0 to the *watch promotion* attribute. This will result in a larger value for both conditional probabilities. However, this is not a problem, because both probability values are equally affected.

5.4.4 Numeric Data

Numeric data can be dealt with in a similar manner provided that the probability density function representing the distribution of the data is known. If a particular numerical attribute is normally distributed, we use the standard probability density function shown in Equation 5.16.

$$f(x) = 1/(\sqrt{2\pi}\sigma)e^{-(x-\mu)^2/2\sigma^2} \tag{5.16}$$

where
e = the exponential function
μ = the class mean for the given numerical attribute
σ = the class standard deviation for the attribute
x = the attribute value

Although this equation looks quite complicated, it is very easy to apply. To demonstrate, consider the data in Table 5.5. This table displays the data in Table 5.3 with an added column containing numerical attribute *age*.

Let's use this new information to compute the conditional probabilities for the *male* and *female* classes for the following instance.

Magazine Promotion = Yes

Watch Promotion = Yes

Life Insurance Promotion = No

Credit Card Insurance = No

Age = 45

TABLE 5.5 Addition of Attribute Age to the Bayes Classifier Dataset

Magazine Promotion	Watch Promotion	Life Insurance Promotion	Credit Card Insurance	Age	Gender
Yes	No	No	No	45	Male
Yes	Yes	Yes	Yes	40	Female
No	No	No	No	42	Male
Yes	Yes	Yes	Yes	30	Male
Yes	No	Yes	No	38	Female
No	No	No	No	55	Female
Yes	Yes	Yes	Yes	35	Male
No	No	No	No	27	Male
Yes	No	No	No	43	Male
Yes	Yes	Yes	No	41	Female

Gender = ?

For the overall conditional probabilities, we have:

$P(E \mid gender = male) = (4/6)\,(2/6)\,(4/6)\,(4/6)\,[P(age = 45 \mid gender = male)]$

$P(E \mid gender = female) = (3/4)\,(2/4)\,(1/4)\,(3/4)\,[P(age = 45 \mid gender = female)]$

To determine the conditional probability for *age* given *gender = male*, we assume *age* to be normally distributed and apply the probability density function. We use the data in Table 5.3 to find the mean and standard deviation scores. For the class *gender = male*, we have: $\sigma = 7.69$, $\mu = 37.00$, and $x = 45$. Therefore, the probability that *age = 45* given *gender = male* is computed as:

$$P(age = 45 \mid gender = male) = 1/(\sqrt{2\pi}\,7.69)e^{-(45-37.00)^2/2(7.69)^2}$$

Making the computation, we have:

$P(age = 45 \mid gender = male) \cong 0.030$

To determine the conditional probability for *age* given *gender = female*, we substitute $\sigma = 7.77$, $\mu = 43.50$, and $x = 45$. Specifically,

$$P(age = 45 \mid gender = female) = 1/(\sqrt{2\pi}\,7.77)e^{-(45-43.50)^2/2(7.77)^2}$$

Making the computation, we have:

$P(age = 45 \mid gender = female) \cong 0.050$

We can now determine the overall conditional probability values:

$$P(E \mid gender = male) = (4/6)\ (2/6)\ (4/6)\ (4/6)\ (0.030) \cong .003$$

$$P(E \mid gender = female) = (3/4)\ (2/4)\ (1/4)\ (3/4)\ (0.050) \cong .004$$

Finally, applying Equation 5.1 we have:

$$P(gender = male \mid E) \cong (0.003)\ (0.60)/P(E) \cong 0.0018/P(E)$$

$$P(gender = female \mid E) \cong (0.004)\ (0.40)/P(E) \cong 0.0016/P(E)$$

Once again, we ignore $P(E)$ and conclude that the instance belongs to the *male* class.

5.4.5 Experimenting with Naïve Bayes

Several packages offer implementations of the naïve Bayes classifier. Here we focus on the *naïveBayes* function contained within package *e1071*—available from the CRAN library.

For our example, we use the dataset described in the box titled *Credit Card Screening Dataset*. The dataset contains information about 690 individuals who applied for a credit card. The data includes 15 input attributes and 1 output attribute indicating whether an individual credit card application was accepted (+) or rejected (–). Naïve Bayes is a good choice for analyzing this data as we don't have to concern ourselves with preprocessing instances with missing data. Also, as Bayes is a probability-based classifier, we have the option of numerically displaying the probabilities associated with each classification. This is a positive with our example as it is likely to be advantageous for a credit card company to deny cards to some individuals who would not default if they are in return able to accurately deny individuals who will default. Knowing the probability associated with each classification allows us to move the *inclusion bar* up or down the list of probabilities as we see fit. One disadvantage is the lack of ability to determine those attributes best suited for building a predictive model. We will use one of the techniques from Chapter 4 to help with this issue.

Here is our goal:

> *Build a predictive model using the instances of the credit card screening dataset able to predict which future credit card applicants to accept and which to reject. Benchmarks for predictive accuracy with previously unseen data include an overall model correctness of 85%, a 10% maximum for incorrect denials, and a 5% maximum for incorrect accepts.*

We provide the partial output of four experiments—listed in Scripts 5.10–5.13—designed to meet our aforementioned goals. All of these scripts reside in the file *Script 5.10–5.13 Bayes classifier.R* and as with all scripts are part of your supplementary materials. Let's look at the first script.

Script 5.10 shows the statements and edited output for our first attempt at building a model able to meet our goals. The *removeNAS* function from Chapter 4 is called here to

THE CREDIT CARD SCREENING DATASET

The Credit Card Screening dataset contains information about 690 individuals who applied for a credit card. The dataset includes 15 input attributes and 1 output attribute indicating whether an individual credit card application was accepted (+) or rejected (−). All input attribute names and values have been changed to meaningless symbols to protect confidentiality of the data. The original dataset was submitted by Ross Quinlan to the UCI machine learning dataset repository—http://archive.ics.uci.edu/ml/index.php.

The dataset is interesting for several reasons. First, the instances represent real data about credit card applications. Second, the dataset offers a nice mix of categorical and numeric attributes. Third, 5% of the dataset records contain one or more pieces of missing information. Finally, as the attributes and values are without semantic meaning, we cannot introduce biases about which attributes we believe to be important.

verify that missing records exist. Your output will list the 24 instances having one or several NA values. We do not use the modified file returned by *removeNAS* as naïve Bayes simply ignores NA attribute values.

The confusion matrix displayed in Script 5.10 shows a 78.7% overall correctness; 35% of the positive class instances had their credit card application rejected. Also, 10.8% (14/130) of those who should have been rejected were offered a credit card. We must do better!

Script 5.11 attempts to improve our results with attribute evaluation. Specifically, *GainRatioAttributeEval* is applied to all of the data to determine the worth of individual attributes relative to the output class. This function is part of the RWeka package (Hornik et al., 2009) available through the CRAN library. Weka, shorthand for *Waikato Environment for Knowledge Analysis*, is a standalone multipurpose tool for data mining and analysis. Weka was developed at the University of Waikato in New Zealand and is publicly available. If you are familiar with Weka, you will be pleased to know that the RWeka package supports most of the classifiers, clustering algorithms, and preprocessing tools within Weka.

The call to *GainRatioAttributeEval* follows the general format used by most of R's machine learning tools. *GainRatioAttributeEval* returns a list that associates a "goodness score" with each attribute. Attribute interactions are not taken into account.

Next, we sort the list to see attribute *nine* as the best scoring input attribute with *ten*, *eleven*, and *fifteen* coming in a distant second, third, and fourth. For the experiment, we eliminated all but the four top scoring input attributes. The resultant confusion matrix shows a decrease in false positives but overall model performance is wanting.

The third experiment (Script 5.12) uses only the top two scoring input attributes. False rejects drop to 10% and overall accuracy increases to 84.78%. However, the 0.19 false positive rate doesn't come close to meeting our stated goal.

Script 5.10 Bayes Classifier Creditscreening Data

```
> library(e1071)
> library(RWeka)


> PREPROCESS DATA
> # Locate but do not delete missing items.
> x <- removeNAS(creditScreening)
  "number deleted"
        24


> set.seed(100)
> credit.data <- creditScreening
> index <- sample(1:nrow(credit.data), 2/3*nrow(credit.data))
> credit.train <- credit.data[index,]
> credit.test <-  credit.data[-index,]


> # CREATE THE MODEL
> credit.Bayes<-naiveBayes(class ~ .,laplace =1, data= credit.train)
> summary(credit.Bayes)

          Length Class  Mode
apriori   2      table  numeric
tables    15     -none- list
levels    2      -none- character
isnumeric 15     -none- logical
call      4      -none- call


> # TEST & CREATE THE CONFUSION MATRIX

> credit.pred <-predict(credit.Bayes, credit.test)
> credit.perf<- table(credit.test$class,
+                     credit.pred, dnn=c("actual", "Predicted"))
> credit.perf


      Predicted
actual   -   +
    - 116  14
    +  35  65


> confusionP(credit.perf)
  Correct= 181
 Incorrect= 49
 Accuracy = 78.7 %
```

Script 5.11 Bayes Classifier: Attribute Selection

```
> # PREPROCESSING

> set.seed(100)
> best <-GainRatioAttributeEval(class ~ ., data=creditScreening)
> best <- sort(best,decreasing = TRUE)

> round(best,3)
   nine    ten   eleven fifteen eight three fourteen four five   six
  0.452  0.161   0.152   0.143  0.114 0.046   0.042 0.033 0.033 0.030
   two   seven thirteen  one   twelve
 0.029  0.029   0.019   0.003   0.001

> credit.Bayes<-naiveBayes(class ~ nine + ten + eleven
+                          + fifteen,laplace = 1,
+                          data= credit.train,type = "class")

> # CREATE CONFUSION MATRIX
> credit.pred <-predict(credit.Bayes, credit.test)
> credit.perf<- table(credit.test$class, credit.pred,
+          dnn=c("actual", "Predicted"))
> credit.perf

      Predicted
actual   -    +
    - 119   11
    +  44   56

> confusionP(credit.perf)

  Correct= 175
Incorrect= 55
Accuracy = 76.09 %
```

Script 5.12 Bayes Classifier: Two Input Attributes

```
> set.seed(100)

> # CREATE THE MODEL
> credit.Bayes<-naiveBayes(class ~ nine + ten, laplace = 1,
+                          data= credit.train)

      Predicted
actual   -    +
    - 105   25
```

```
  +   10   90
```

```
> confusionP(credit.perf)
  Correct= 195
Incorrect= 35
Accuracy = 84.78 %
```

Script 5.13 Bayes Classifier: Numeric Output

```
> set.seed(100)
> # CREATE THE MODEL
> credit.Bayes<-naiveBayes(class ~ nine + ten, laplace =1,
+                          data= credit.train)

> credit.pred <-predict(credit.Bayes, credit.test,laplace=1,
type="raw")
> credit.pred <- cbind(credit.test$class,credit.pred)
> head(credit.pred)


            -            +
[1,]  2  0.38359702  0.6164030
[2,]  2  0.38359702  0.6164030
[3,]  2  0.38359702  0.6164030

> credit.df <- data.frame(credit.pred)
> colnames(credit.df) <- c("class","No","yes") # change column
names
> credit.ordered <- credit.df[order(credit.df$yes, decreasing =
TRUE),]
> head(credit.ordered)

    class       No      yes
5       2  0.08439337  0.9156066
6       2  0.08439337  0.9156066
8       2  0.08439337  0.9156066
....... .
```

For the final experiment, we directed the *predict* function to give us the *raw* posterior probabilities computed by the classifier. The idea is to examine individual probability values in order to find a split point that will enable us to remove some of the false positive classifications without losing too many true positives. To investigate this possibility, we attach the computed probabilities to their actual class with *cbind*. The result is converted to a data frame, the + and – headings are replaced with *yes* and *no*, and the *order* function sorts the probabilities from high to low based on the *yes* column. Script 5.13 lists the first three sorted values. All three instances belong to the *yes* class (2). If you scroll the entire list of sorted probability values, you will see the first 79 instances have a predicted probability of 0.916. Seven of these instances are false accepts. The 80th instance has an associated probability of 0.616 as do the next 35 instances. The 116th instance represents the break between the two classes. Given this, even if we limit issuing credit cards to the individuals with an associated probability above 90%, we still do not pass the 5% false positive test. The end-of-chapter exercises provide an opportunity for additional work with this dataset.

5.5 CHAPTER SUMMARY

A favorite statistical technique for estimation and prediction problems is linear regression. Linear regression attempts to model the variation in a numeric dependent variable as a linear combination of one or more numeric independent variables. Linear regression is an appropriate strategy when the relationship between the dependent and independent variables is nearly linear. Linear regression is a poor choice when the outcome is binary. The problem lies in the fact that the value restriction placed on the dependent variable is not observed by the regression equation. That is because linear regression produces a straight-line function, and values of the dependent variable are unbounded in both the positive and negative directions. For the two-outcome case, logistic regression is a better choice. Logistic regression is a nonlinear regression technique that associates a conditional probability value with each data instance. With possible outcomes 0 or 1, the probability $p(y = 1 \mid \mathbf{x})$ represents the conditional probability that the data instance with attribute vector $\mathbf{x}$ is part of the class associated with outcome 1. Likewise, $1 - p(y = 1 \mid \mathbf{x})$ is the probability that the instance is part of the class associated with outcome 0.

The Naïve Bayes classifier offers a simple yet powerful supervised classification technique. The model assumes all input attributes to be of equal importance and independent of one another. Even though these assumptions are likely to be false, the Bayes classifier still works quite well in practice. Naïve Bayes can be applied to datasets containing both categorical and numeric data. Also, unlike many statistical classifiers, the Bayes classifier can be applied to datasets containing a wealth of missing items. Our chances of doing well with Naïve Bayes oftentimes significantly improves when attribute selection takes place prior to model building.

5.6 KEY TERMS

- *A priori probability.* The probability a hypothesis is true lacking evidence to support or reject the hypothesis.

- *Bayes classifier.* A supervised learning approach that classifies new instances by using Bayes theorem.

- *Bayes theorem.* The probability of a hypothesis given some evidence is equal to the probability of the evidence given the hypothesis, times the probability of the hypothesis, divided by the probability of the evidence.

- *Conditional probability.* The conditional probability of evidence E given hypothesis H denoted by $P(E \mid H)$ is the probability E is true given H is true.

- *Degrees of freedom.* The number of values in the calculation of a statistic that are free to vary.

- *Density plot.* A graph that allows us to visualize the distribution of data over a continuous interval.

- *False positive rate (FP rate).* FP rate is computed as the total number of instances incorrectly classified with a specific class divided by the total number of instances that are not part of the class.

- *Laplace smoothing.* A correction parameter where one is added to each attribute value count in order to avoid zero probabilities.

- *Least-squares criterion.* The least-squares criterion minimizes the sum of squared differences between actual and predicted output values.

- *Linear regression.* A statistical technique that models the variation in a numeric-dependent variable as a linear combination of one or several independent variables.

- *Logistic regression.* A nonlinear regression technique for problems having a binary outcome. The equation limits the values of the output attribute to values between 0 and 1. This allows output values to represent a probability of class membership.

- *Logit.* The natural logarithm of the odds ratio $p(y = 1 \mid \mathbf{x})/[1 - p(y = 1 \mid \mathbf{x})]$. $p(y = 1 \mid \mathbf{x})$ is the conditional probability that the value of the linear regression equation determined by feature vector $\mathbf{x}$ is 1.

- *Residual.* An error computed as the difference between actual and computed output.

- *Residual standard error.* The average error in predicting output values. It is computed as the square root of the sum of squares of all residuals divided by the degrees of freedom.

- *Significance level.* The probability at which we are willing to reject that the result of an experiment is due purely to chance.

- *Simple linear regression.* A regression equation with a single independent variable.

- *Slope–intercept form.* A linear equation of the form $y = ax + b$ where a is the slope of the line and b is the y intercept.

- *True positive rate (TP rate).* TPR is computed by dividing the number of instances correctly classified within a class by the total number of instances actually belonging to the class.

EXERCISES

Review Questions

1. Differentiate between the following:

    a. Simple and multiple linear regression

    b. Linear and logistic regression

    c. A priori, conditional, posterior probability

Experimenting with R

1. Execute Script 5.1 and type *attributes(my.slr)*. This function lists the attributes associated with an object. Type *my.slr$* followed by an attribute name to list the output for five of the attributes.

2. Repeat Exercise (1) but use *hr32* as the sole input variable.

3. Repeat Exercise (1) using the OFCM.csv dataset with *oswest.in* as the sole input variable and *oswest.out* as the response variable.

4. Create a multiple linear regression model for the gamma-ray burst dataset as follows:

    a. Build a model to determine *t90* using *fl*, *hr32*, and *t50*.

    b. Use the *anova* function to determine the input attribute contributing the least to the sum of squared error reduction.

    c. Use the *update* function to remove the least effective input attribute from the original model.

    d. Use the *anova* function to determine if the squared error reduction seen by the model created using the *update* function is significant.

5. Repeat the experiment given in Script 5.5 but remove oswest.out and add LE.wt.out as the response variable. Comment on the significance of the regression equation coefficients and report on the statistics computed by *mmr.stats*. Are the input variables able to accurately predict LE.wt.out?

6. Import the file containing Scripts 5.8 & 5.9 into RStudio. Locate the line within Script 5.9 that appears as follows: #p.card <- ifelse(p.card > 0.5,2,1). Remove the comment and execute both Scripts 5.8 and 5.9. You will notice that the inclusion of the ifelse takes the *curves* out of the ROC curve. This is the case as the decimal values within p.card are now all 1's and 2's.

    a. Change the 0.5 cutoff for the *sick* class to 0.4, then to 0.3, 0.2, and finally 0.1 each time recording the value for *auc*.

b. Repeat this process but use 0.6, 0.7, 0.8, and lastly 0.9 for the cutoff value.

c. Describe in detail the effect these changes have on the value of *auc*.

7. Repeat Script 5.11 but use attribute *nine* as the lone input attribute. Next, build a second model by replacing attribute *nine* with attribute *twelve*. Build a final model using both attributes for input. Which model shows the best test set accuracy?

8. Use Script 5.8 as a template to build and test two logistic regression models. Build and test one model using only those attributes shown to have significant coefficients. Build and test a second model using only those attributes in Script 5.8 having a significant deviance score. Compare the test set accuracy of the models to each other and to the original model.

9. Read the documentation for the Spam dataset. The documentation is written in the *description* sheet of Spam.xlsx. Write a script to compare R's logistic regression function with naïve Bayes using the .csv version of the dataset. Here is the procedure:

a. Import the Spam dataset.

b. Use the removeNAS function to delete any instances with one or more missing attribute values.

c. Determine the accuracy of each technique using a 2/3 training, 1/3 test set evaluation.

d. Return to preprocess mode and use *GainRatioAttributeEval* to eliminate all but the six best input attributes. Repeat *c*.

e. Report your findings.

Computational Questions

1. Use the data contained in Table 5.3 to fill in the counts and probabilities in the following table. The output attribute is *life insurance promotion*.

	Magazine Promotion		Watch Promotion		Credit Insurance			Gender	
	Yes	No	Yes	No	Yes	No		Yes	No
Life insurance promotion									
Yes							Male		
No							Female		
Ratio: yes/total							Ratio: male/total		
Ratio: no/total							Ratio: female/total		

   a. Use the completed table together with the naïve Bayes classifier to determine the value of *life insurance promotion* for the following instance:

      Magazine Promotion = Yes

      Watch Promotion = Yes

      Credit Card Insurance = No

      Gender = Female

      Life Insurance Promotion = ?

   b. Repeat part a, but assume that the *gender* of the customer is unknown.

   c. Repeat part a, but use Equation 5.15 with $k = 1$ and $p = 0.5$ to determine the value of *life insurance promotion*.

2. Consider the confusion matrix below where *Yes* represents the positive class.

		Predicted Class	
		Yes	No
Actual class	Yes	30	10
	No	10	70

   a. Compute the overall classification accuracy.

   b. Compute the true positive rate.

   c. Compute the false positive rate.

Installed Packages and Functions

Package Name	Function(s)
base/stats	abline, anova, aov,c, cbind, cor, data.frame, df.residual, factor, fitted, glm,head, ifelse, library, lm, order, plot, predict, residuals, round, sample, set.seed, step, summary, table, update
Car	densityPlot, scatterplotMatrix
Caret	train, trainControl
e1071	naiveBayes
graphics	boxplot
RWeka	GainRatioAttributeEval
ROCR	performance, prediction

## CHAPTER 6

Tree-Based Methods

In This Chapter

- A Decision Tree Algorithm

- Building Decision Trees

- Ensemble Techniques

- Regression Trees

CHAPTER 5 FOCUSED ON supervised statistical machine learning techniques where certain assumptions such as normality and data independence are required in order to validate results. We now turn our attention to supervised machine learning algorithms that make no assumptions about the nature of the data to be analyzed.

In Section 6.1, our focus is on standard methods for creating decision trees. In Section 6.2, we apply a commerical version of Quinlan's C4.5 decision tree algorithm to customer churn data. Section 6.3 presents *rpart*, a tree-based learning model for constructing decision and regression trees. In Section 6.4, we introduce the RWeka machine learning package and apply its J48 decision tree algorithm to the customer churn data. The focus of Section 6.5 is on ensemble techniques for improving performance. Building regression trees with *rpart* is the topic of the final section of this chapter.

6.1 A DECISION TREE ALGORITHM

Decision trees are a popular structure for supervised learning. Countless articles have been written about successful applications of decision tree models to real-world problems. We introduced the C4.5 decision tree model in Chapter 1. In this section, we take a closer look at the algorithm used by C4.5 for building decision trees.

6.1.1 An Algorithm for Building Decision Trees

Decision trees are constructed using only those attributes best able to differentiate the concepts to be learned. A decision tree is built by initially selecting a subset of instances from
a training set. This subset is then used by the algorithm to construct a decision tree. The
remaining training set instances test the accuracy of the constructed tree. If the decision
tree classifies the instances correctly, the procedure terminates. If an instance is incorrectly
classified, the instance is added to the selected subset of training instances and a new tree
is constructed. This process continues until a tree that correctly classifies all nonselected
instances is created or the decision tree is built from the entire training set. We offer a simplified version of the algorithm that employs the entire set of training instances to build a
decision tree. The steps of the algorithm are as follows:

1. Let *T* be the set of training instances.
2. Choose an attribute that best differentiates the instances contained in *T*.
3. Create a tree node whose value is the chosen attribute. Create child links from this
   node where each link represents a unique value for the chosen attribute. Use the child
   link values to further subdivide the instances into subclasses.
4. For each subclass created in step 3:
   a. If the instances in the subclass satisfy predefined criteria or if the set of remaining
      attribute choices for this path of the tree is null, specify the classification for new
      instances following this decision path.
   b. If the subclass does not satisfy the predefined criteria and there is at least one attri-
      bute to further subdivide the path of the tree, let *T* be the current set of subclass
      instances and return to step 2.

The attribute choices made when building a decision tree determine the size of the constructed tree. A main goal is to minimize the number of tree levels and tree nodes, thereby
maximizing data generalization. With the help of our credit card promotion dataset, the
next section details the attribute selection method used by C4.5.

6.1.2 C4.5 Attribute Selection

C4.5 uses a measure taken from *information theory* to help with the attribute selection
process. At each choice point in the tree, C4.5 computes a *gain ratio* for all available attributes. The attribute with the largest value for this ratio is selected to split the data. Here is
the equation to compute the gain ratio for attribute *A*.

$$GainRatio(A) = Gain(A) / Split\ Info(A)$$

For a set of *I* instances, the equation for computing *Gain(A)* is given as:

$$Gain(A) = Info(I) - Info(I, A)$$

where *Info(I)* is the information contained in the currently examined set of instances, and *Info(I,A)* is the information after partitioning the instances in *I* according to the possible outcomes for attribute *A*.

The equations for computing *Info(I)*, *Info(I,A)*, and *Split Info(A)* are straightforward. For *n* possible classes, the equation for computing *Info(I)* is

$$Info(I) = -\sum_{i=1}^{n} \frac{\#\,in\,class\,i}{\#\,in\,I} \log_2 \left(\frac{\#\,in\,class\,i}{\#\,in\,I} \right)$$

After *I* is partitioned into *k* outcomes, *Info(I,A)* is computed as

$$Info(I,A) = \sum_{j=1}^{k} \frac{\#\,in\,class\,j}{\#\,in\,I}\, info(class\,j)$$

Finally, Split *Info(A)* normalizes the gain computation to eliminate a bias for attribute choices with many outcomes. Without using the value for Split Info, attributes with unique values for each instance will always be selected. Once again, for *k* possible outcomes

$$Split\,Info(A) = -\sum_{j=1}^{k} \frac{\#\,in\,class\,j}{\#\,in\,I} \log_2 \left(\frac{\#\,in\,class\,j}{\#\,in\,I} \right)$$

Let's use the data given in Table 6.1 to see how the equations are applied. For our example, we designate *life insurance promotion* as the output attribute. For illustrative purposes, we apply the equations to the entire set of training data. We wish to develop a

TABLE 6.1   The Credit Card Promotion Dataset

Id.	Income Range	Life Insurance Promotion	Credit Card Insurance	Gender	Age
1	40–50 K	No	No	Male	45
2	30–40 K	Yes	No	Female	40
3	40–50 K	No	No	Male	42
4	30–40 K	Yes	Yes	Male	43
5	50–60 K	Yes	No	Female	38
6	20–30 K	No	No	Female	55
7	30–40 K	Yes	Yes	Male	35
8	20–30 K	No	No	Male	27
9	30–40 K	No	No	Male	43
10	30–40 K	Yes	No	Female	41
11	40–50 K	Yes	No	Female	43
12	20–30 K	Yes	No	Male	29
13	50–60 K	Yes	No	Female	39
14	40–50 K	No	No	Male	55
15	20–30 K	Yes	Yes	Female	19

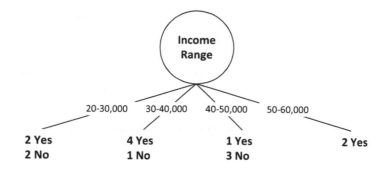

FIGURE 6.1   A partial decision tree with root node = income range.

predictive model. Therefore, the input attributes are limited to *income range, credit card insurance, gender,* and *age.* The purpose of the attribute designated as *Id* is to reference items within the table.

Figure 6.1 shows the results of a top-level split on the attribute *income range.* The total *yes* and *no* counts for the output attribute *life insurance promotion* are shown at the bottom of each branch. Although this attribute is not a good choice for splitting the table data, it is a good selection for demonstrating the equations described above.

We first compute the value for *Info(I).* Since we are trying to determine the top-level node of the tree, *I* contains all 15 table instances. Given that *life insurance promotion* contains nine *yes* values and six *no* values, the computation is

$$Info(I) = -[9/15\log_2 9/15 + 6/15\log_2 6/15] = 0.97095$$

Income range has four possible outcomes, the computation of *Info(I, Income range)* is

$$Info(I, Income\ Range) = 4/15 Info(20-30,000) + 5/15 Info(30-40,000)$$
$$+ 4/15 Info(40-50,000) + 2/15 Info(50-60,000)$$
$$= 0.72365$$

where,

$$Info(20-30,000) = -[2/4\log_2 2/4 + 2/4\log_2 2/4];$$

$$Info(30-40,000) = -[4/5\log_2 4/5 + 1/5\log_2 1/5];$$

$$Info(40-50,000) = -[3/4\log_2 3/4 + 1/4\log_2 1/4];$$

$$Info(50-60,000) = -[2/2\log_2 2/2];$$

also,

$$SplitInfo(Income\ Range) = -\left[4/15\log_2 4/15 + 5/15\log_2 5/15 +\right.$$

$$4/15\log_2 4/15 + 2/15\log_2/15\Big]$$

$$\approx 1.93291$$

We now compute the gain as

$$gain(Income\ Range) = Info(I) - Info(I, Income\ Range)$$

$$\approx 0.97905 - 0.72193 = 0.25712$$

Finally,

$$Gain\ Ratio(Income\ Range) = Gain(Income\ Range) / Split\ Info(Income\ Range)$$

$$\approx 0.25712 / 1.93291 = 0.13302$$

Scores for the categorical attributes *credit card insurance* and *gender* are computed in a similar manner. The question now becomes how do we apply the equations in order to obtain a score for the numerical attribute *age*? The answer is to discretize the data by sorting numeric values and computing a gain ratio score for each possible binary split point. For our example, the ages are first sorted as

19	27	29	35	38	39	40	41	42	43	43	43	45	55	55
Y	N	Y	Y	Y	Y	Y	Y	N	Y	Y	N	N	N	N

Next, we compute a score for each possible split point. That is, the score for a binary split between 19 and 27 is computed as is the score for a split between 27 and 29. This process continues until a score for the split between 45 and 55 is obtained. In this way, each split point is treated as a separate attribute with two values. Finally, by making computations for *income range, credit card insurance, gender,* and *age,* we find that *credit card insurance* has the best gain ratio score of 3.610. Figure 6.2 displays the partial decision tree with *credit card insurance* as the root node.

This takes us to Step 4. (For each subclass…) of our decision tree algorithm where a decision must be made for each of the two branches of the partial tree. First, consider the branch showing *Credit Card Insurance = Yes.* All three instances following this path have *Life Insurance Promotion = Yes.* Therefore, *Step 4.a* applies and the algorithm terminates this path with the label *Life Insurance Promotion = Yes.*

The path *Credit Card Insurance = No* shows six instances with *Life Insurance Promotion =Yes* and six with *Life Insurance Promotion = No.* Therefore, *Step 4.b* applies and the algorithm returns to Step 2 where attribute selection is applied to the remaining

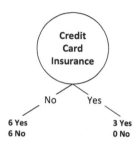

FIGURE 6.2   A partial decision tree with root node = credit card insurance.

12 instances. *Gender* is the winning attribute resulting in the decision tree displayed in Figure 6.3. Following the branch, *Gender = Female* shows five of the six female customers said *yes* to the promotion. Therefore, *Step 4.a* applies and the path terminates with the label *Life Insurance Promotion = Yes*. In the same manner, the branch *Gender = Male* shows five of the six male customers said *no* to the promotion. Therefore, *Step 4.a* applies and the path terminates with the label *Life Insurance Promotion = No*. It is instructive to point out that both paths for *Gender* misclassify one instance. As the attributes *age* and *income range* are still available, it is still possible to further subdivide the tree. However, it is highly likely that one of the predefined criteria for non-termination of a specific branch would require more than one remaining misclassified instance. Implementations of decision tree algorithms allow the user to decide on the tradeoff between tree generality and training data classification accuracy.

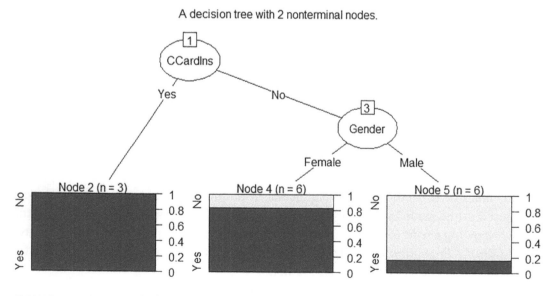

FIGURE 6.3   A two-node decision tree for the credit card promotion dataset.

6.1.3 Other Methods for Building Decision Trees

We just described the basics of C4.5 — Quinlan's most recent noncommercial decision tree building algorithm. However, several other algorithms for building decision trees exist. ID3 (Quinlan, 1986) has been studied extensively and is the precursor to C4.5.

CART (Breiman et al., 1984) — the acronym for a class of tree-based methods known as classification and regression trees — is of particular interest as several commercial products implement variations of the original algorithm. CART is very similar to C4.5, but there are several differences. One notable difference is that CART always performs binary splits on the data regardless of whether attributes are categorical or numeric. Another significant difference is the measure used for attribute selection. Rather than using C4.5's information gain approach, the default measurement used by CART at each choice point is an impurity score known as the *Gini*. Specifically,

$$Gini = 1 - \sum_{n=1}^{2} \left(p_i(n) \right)^2$$

where $p_i(n)$ is the fraction of instances belonging to class i at node n. The attribute chosen for the next split of the decision tree is the attribute split resulting in the smallest *Gini* computation. As you can see, when all instances at node n are of the same class, the summation is 1, thereby giving a Gini impurity score of 0.

A third difference is that CART invokes test data to help prune and therefore generalize a created binary tree, whereas C4.5 uses only training data to create a final tree structure. Lastly, CART was the first system to introduce *regression trees*. Regression trees take the form of decision trees where the leaf nodes are numerical rather than categorical values.

CHAID (Kass, 1980) — Chi-square automatic interaction detection — is a third decision tree building algorithm of interest found in commercial statistical packages such as SAS and SPSS. CHAID differs from C4.5 and CART in that it is limited to working with categorical attributes. CHAID has a statistical flavor as it uses the X^2 statistical test of significance to determine candidate attributes for building the decision tree. In the sections that follow, we will experiment with R's versions of C4.5, C5.0, and an implementation of CART found in the *rpart package*. Let's begin!

6.2 BUILDING DECISION TREES: C5.0

Our first look at building decision trees starts with C5.0, an enhanced version of C4.5. C5.0 is based on Quinlan's commerically available model but is lacking some features of the full commercial version. The decision tree displayed in Figure 6.3 was constructed using C5.0. The statements used to build the tree in Figure 6.3 as well as relevant output is listed in Script 6.1. Before executing the script, you must install the C50 package. Also, be sure to have *ccpromo.csv* imported into RStudio. Let's take a closer look at Script 6.1.

Script 6.1 A C5.0 Decision Tree for the Credit Card Promotion Dataset

```
> library(C50)

> ccp.C50 <- C5.0(LifeInsPromo ~ ., data = ccpromo,
+                 control=C5.0Control(minCases=2))
> summary(ccp.C50)

Call:
C5.0.formula(formula = LifeInsPromo ~ ., data =
 ccpromo, control = C5.0Control(minCases = 2))

Decision tree:

CCardIns = Yes: Yes (3)
CCardIns = No:
:...Gender = Female: Yes (6/1)
    Gender = Male: No (6/1)

Evaluation on training data (15 cases):

    Decision Tree
  ----------------
   Size       Errors

     3      2(13.3%)    <<

    (a)    (b)      <-classified as
    ----   ----
     5      1      (a): class No
     1      8      (b): class Yes

Attribute usage:

100.00% CCardIns
80.00% Gender

> plot(ccp.C50, main = "A decision tree with two nonterminal
nodes.")
```

6.2.1 A Decision Tree for Credit Card Promotions

The call to C5.0 shows *LifeInsPromo* as the output attribute and specifies that all input attributes are to be considered for building the tree. *minCases* instructs the algorithm to consider splitting any tree node having at least two instances remaining in the current path. The *summary* function states the call to C5.0 afterwhich we see statements defining the tree. Terminal nodes provide a classification, indicate the number of instances following the given path, and tell us how many of these instances have been incorrectly classified. For example, the branches leading to *Gender = Female: Yes (6/1)* tell us that six instances follow this path with one (instance #6) representing a misclassification. The misclassified instance satisfies CCardIns = *No* and Gender = *Female* but shows a value of *No* for LifeInsPromo.

Next, we see an evaluation on the training data in the form of a confusion matrix. The matrix shows that two training instances are misclassified by the tree. The list under *Attribute usage* states those attributes used to build the corresponding decision tree. Attributes at the top of the list are most relevant for predicting outcome. Lastly, the *plot* function gives us the tree shown in Figure 6.3.

Before moving ahead to a more interesting example, it is worth your time to check out the options available with C5.0. Simply looking at the documentation for C5.0 as given in the C50 package is very instructive. Of primary interest is that C5.0 has an alternative call format. To use this format in Script 6.1, we would write

```
ccp.C50 <- C5.0(ccpromo[-4],ccpromo$LifeInsPromo, control=C5.0Control(minCases=2))
```

With this format, we have the input data listed first. Notice the output attribute (column 4) has been eliminated from the data. The output attribute is listed next followed by the C5.0 control statement.

A more general declaration is

```
my.Model <- C5.0(trainData, class, trials=1, costs=null,control=C5.0Control( ))
```

where *trials, costs, and control* are optional.

A value greater than 1 for *trials* signals a multiple model approach known as *boosting* (see Section 6.5). *Costs* allows us to associate variable costs with different types of errors. Let's take a look at a dataset where experimenting with the trials parameter as well as *minCases* may be worth our time and effort.

6.2.2 Data for Simulating Customer Churn

Two artificial datasets originally contained in the MLC++ machine learning library simulate telecommunications customer churn. The datasets hold 19 input attributes, most of which are numeric. The output attribute is *churn* with possible values *yes* or *no*. The training data includes 3333 customer instances and is contained in *churnTrain*. The 1667 instance test set is found in *churnTest*.

Our goal is to build a predictive model with C5.0 that is able to determine whether a given customer is likely to churn. Correctly identifying likely churn candidates is of primary importance as it allows the opportunity to offer incentives to discourage these customers from dropping their service. To follow our explanation, import both datasets and load *Script 6.2*—all part of your supplementary materials—into RStudio!

6.2.3 Predicting Customer Churn with C5.0

Script 6.2 lists the most relevant statements and output for our training/test set experiment with C5.0. Initially, we used the default setting of 2 for *minCases*. This resulted in a 27 terminal node tree showing a test set accuracy of 94.72% with 78 misclassified churners. The model defined in Script 6.2 with *mincases* = 50 gives the 8 terminal node tree displayed in Figure 6.4. The test set accuracy of the smaller tree remains high at 93.22% with 83 misclassified churners. The question as to whether a better understood, more efficient, but less accurate tree is a better choice is best answered by determining if observed differences are statistically significant. Testing for statistically significant differences when comparing competing models is a central topic of Chapter 9.

Lastly, Script 6.25 (not shown here) can be found in the scripts.zip file for Chapter 6. The script shows how the technique first illustrated in Script 5.13 is used to obtain predictions as probability values. The script also illustrates how to list the specific instances with incorrect classifications. Knowing which instances are incorrectly classified can often help with the development of an improved model. Next, we turn our attention to a well-known implementation of CART.

Script 6.2 Predicting Customer Churn: C5.0

```
> #PREPROCESSING
> library(C50)
> set.seed(100)

> # CREATE THE DECISION TREE
> churn.C50 <-C5.0(churn ~ ., data=churnTrain, trials=1,
+                 control=C5.0Control(minCases=50))
> churn.C50

Tree size: 8


    Attribute usage:

    100.00%      total_day_minutes
     93.67%      number_customer_service_calls
     86.14%      international_plan
      8.01%      total_intl_calls
      6.48%      total_intl_minutes
      6.33%      voice_mail_plan

> # TEST THE MODEL
> churn.pred <- predict(churn.C50, churnTest,type="class")
> churn.conf <- table(churnTest$churn,churn.pred,
+     dnn=c("Actual","Predicted"))
```

```
> churn.conf

        Predicted

Actual  yes   no
   yes  141   83
   no    30 1413

> confusionP(churn.conf)
  Correct= 1554 Incorrect= 113
  Accuracy = 93.22 %

> # PLOT THE TREE
> plot(churn.C50,main
+    ="A C5.0 decision tree for customer churn with mincases=50.")
```

6.3 BUILDING DECISION TREES: RPART

The *rpart* function in the *rpart package* is a familiar implementation of CART. Let's start with a simple example once again using the credit card promotion data displayed in Table 6.1. Let's see how the decision tree generated by *rpart* matches against the C5.0 decision tree shown in Figure 6.3.

The statements for building the tree as well as relevant output are listed in Script 6.3. Before executing the script, be sure to install the *rpart*, *rpart.plot*, and *partykit* packages available in the CRAN library.

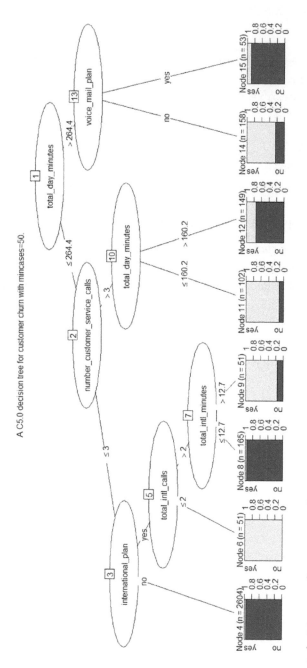

FIGURE 6.4 A decision tree for customer churn: mincases = 50.

Script 6.3 An rpart Decision Tree for Credit Card Promotion Data

```
> library(partykit)
> library(rpart)
> library(rpart.plot)
> set.seed(100)
> ccp.rpart <- rpart(LifeInsPromo ~ ., data = ccpromo,
+                    control=rpart.control(minsplit=3),method="class")
> ccp.rpart

node), split, n, loss, yval, (yprob)
      * denotes terminal node

 1) root 15 6 Yes (0.4000000 0.6000000)
   2) Age>=44 3 0 No (1.0000000 0.0000000) *
   3) Age< 44 12 3 Yes (0.2500000 0.7500000)
     6) Gender=Male 6 3 No (0.5000000 0.5000000)
      12) CCardIns=No 4 1 No (0.7500000 0.2500000) *
      13) CCardIns=Yes 2 0 Yes (0.0000000 1.0000000) *
     7) Gender=Female 6 0 Yes (0.0000000 1.0000000) *

> prp(ccp.rpart,main="Decision Tree with Three Non-Terminal Nodes",
+     type=2,extra=104, fallen.leaves=TRUE,roundint=FALSE)

> plot(as.party(ccp.rpart),
+      "A Decision Tree with three Non-Terminal Nodes")

> # Decision tree rules.
> rpart.rules(ccp.rpart, roundint=FALSE)
 LifeInsPromo

          0.00 when Age >= 44
          0.25 when Age <  44 & Gender is   Male & CCardIns is  No
          1.00 when Age <  44 & Gender is   Male & CCardIns is Yes
          1.00 when Age <  44 & Gender is Female
```

6.3.1 An rpart Decision Tree for Credit Card Promotions

The library statements in Script 6.3 reflect three packages. The *rpart* package is where we find the *rpart* function that implements a version of CART. The *rpart.plot* library contains the *rpart.plot* and *prp* functions for plotting rpart decision trees. It also houses the *rpart.rules* function for mapping an rpart decision tree to a set of rules. The *partykit* package allows us to coerce an *rpart*-created object so it can be printed as an object of type *party*.

The call to *rpart* tells us that all input attributes are to be considered for building the tree. Also, there must be at least three tree node instances before a split is considered. *Method = class* is assumed if the output attribute is a factor which is the case here.

The next few lines give us the statements defining the decision tree. Initially, it can be a challenge to understand the written description. Consider the first line where we have

```
root 15 6 Yes (0.4000000 0.6000000)
```

This line tells us that at the root level we have 15 instances, six of which go against the most common value (nine instances show LifeInsPromo = *Yes*). The parenthesis displays the probability scores for *no* and *yes*. The next two lines give us

```
2) Age>=44 3 0 No (1.0000000 0.0000000) *
3) Age< 44 12 3 Yes (0.2500000 0.7500000)
```

Notice the *Gini* measurement chose *age* rather than *credit card insurance* for the top node of the tree. Line 2 is a terminal path as indicated by the *. It tells us that three instances follow *Age* >=44. The *0* indicates all three instances are *No*. Line 3 shows 12 instances follow the path having *Age* < 44. The common value for LifeInsPromo is *Yes* with 3 instances showing *No*. *Gender* is seen as the next attribute choice where *Gender* = *Female* is a terminal path. *Gender* = *Male* splits on *CCardIns* where both values are terminal.

Graphical pictures of the tree obtained with *prp* (Figure 6. 5) and *plot* (Figure 6.6) are much easier to interpret! Your copy of Script 6.3 also includes *rpart.plot* as a third method. *Prp* and *rpart.plot* offer several options worth investigating. With *prp*, type =2 places split labels beneath each node. *Fallen.leaves* = *TRUE* places all terminal nodes at the same level on the bottom of the tree, and *extra* = *104* includes probability values in the graph. Setting *roundint* to false avoids a warning seen when one or more numeric attributes within the training data are strictly integer. The warning has to do with the possibility of newly presented data having decimal values. Lastly, *rpart.rules* maps the decision tree to a set of rules. The decimal values give the fraction of instances having *LifeInsPromo* = *yes* that satisfy each rule.

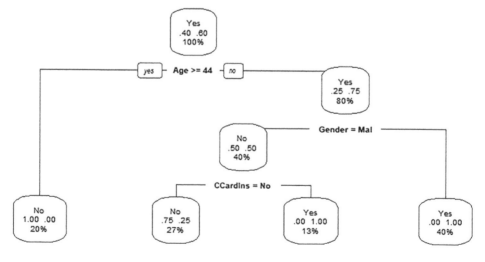

FIGURE 6.5   A three-node decision tree for credit card promotions: prp function.

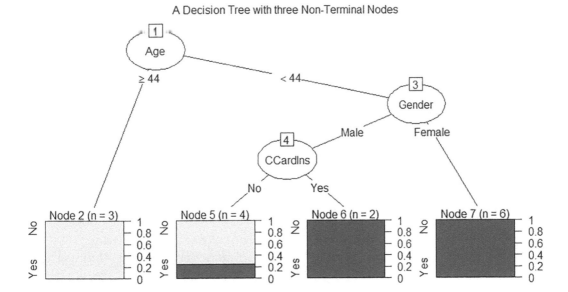

FIGURE 6.6   A three-node decision tree for credit card promotions: partykit.

Given the results of our experiment, we conclude that the trees created by C5.0 and *rpart* for the credit card promotion data are not the same. This is not surprising as each model uses a different attribute selection technique. However, it is worth noting that the Gini scores for splitting *age* at 44 and for *CCardins* are identical as both splits result in one three-instance terminal node.

In the next two sections, we apply *rpart* to the C50 churn data described above. First, we use a training/test set scenario to show how we can determine a best model with the help of the tools provided by *rpart*. After this, we show how *rpart* works in concert with the cross validation functions available with the *caret* package, thereby relieving us of much of the work required for choosing a best model. Once again, our goal is to build a predictive model able to identify likely churners. Let's get churning!

6.3.2 Train and Test rpart: Churn Data

Script 6.4 provides a listing of the most relevant statements and resultant output for our training/test set scenario. By default, *rpart* builds its model using a tenfold cross validation on the training data. The *xval* parameter—currently set at its default value—can be used to modify the number of folds. The tree (not displayed) created by *rpart* using the default settings contains 121 nodes.

When applied to the test data, the model showed an overall accuracy of 92.98% with 63% of all churners correctly identified. As our goal is to identify likely churners, this result is less than optimal. Also, with such a large-sized tree, overfitting is a possibility. Can we do better with a smaller tree? The complexity table helps us answer this question by providing useful error information about trees of various sizes. The complexity table is listed in the model summary but can also be displayed with the statement *churn.rpart$cptable*. Let's take a look at the information in the complexity table displayed in Script 6.4.

Script 6.4 Train/Test Scenario with rpart: Churn Data

```
> #PREPROCESSING
> library(rpart)
> library(rpart.plot)
> library(partykit)
> set.seed(100)


> # CREATE THE DECISION TREE
> churn.rpart <-rpart(churn ~ ., data=churnTrain,
+                      control=rpart.control(xval=10),method ="class")

> churn.pred <- predict(churn.rpart, churnTest,type="class")
> churn.conf <- table(churnTest$churn,churn.pred,dnn=c("Actual",
"Predicted"))
```

```
        Predicted
Actual   yes    no
   yes   141    83
   no     34  1409
```

Accuracy = 92.98 %

```
> # LIST THE CPTABLE PLOT CP VS. X-VAL ERROR
> churn.rpart$cptable
          CP nsplit rel error    xerror       xstd
1 0.08902692      0 1.0000000 1.0000000 0.04207569
2 0.08488613      1 0.9109731 0.9668737 0.04148888
3 0.07867495      2 0.8260870 0.8778468 0.03982814
4 0.05279503      4 0.6687371 0.6811594 0.03565194
5 0.02380952      7 0.4741201 0.4865424 0.03059918
6 0.01759834      9 0.4265010 0.4906832 0.03071920
7 0.01449275     12 0.3685300 0.5031056 0.03107546
8 0.01000000     14 0.3395445 0.4927536 0.03077897
> plotcp(churn.rpart)

> # CREATE AND TEST A NEW MODEL
>    churnNew.rpart <- prune(churn.rpart,cp=0.02381)
> # TEST THE NEW MODEL
> churn.pred <- predict(churnNew.rpart, churnTest,type="class")
> churn.conf <- table(churnTest$churn,churn.pred,dnn=c("Actual",
"Predicted"))

> churn.conf
        Predicted
```

```
Actual   yes    no
   yes   141    83
   no     28  1415


Accuracy = 93.34 %

plot(as.party(churnNew.rpart),
    "A pruned rpart tree for customer churn.")
```

The table is based on tree splits where the entries are listed starting with a tree of zero splits. The complexity parameter (cp) represents the tradeoff between cross-validated relative error rate and tree size as measured by the number of node splits. The *xerror* column gives the cross-validated error based on the training data. The column labeled *xstd* is the standard error of the cross-validated error—the error of an error! Notice that the relative error and cross-validated error are both scaled so the tree with zero splits shows an error of 1.

To better understand the table, we use *plotcp* to plot cp against the cross-validated relative error (Figure 6.7). The graph as well as the complexity table indicates little improvement in relative error after seven node splits. With this information, we use the *prune* function with cp = 0.02381 to create a pruned tree of the desired size. Figure 6.8 shows the top-level nodes of the pruned tree which contains 59 rather than 121 nodes. When tested, the pruned tree shows a slight increase in accuracy but still misclassifies 83 churners. In summary, overall model speed (fewer tree nodes) and accuracy have been improved but there is still work to be done. Let's see if we get a better result with cross validation and the *caret package!*

6.3.3 Cross Validation rpart: Churn Data

Script 5.4 in Chapter 5 introduced you to cross validation with linear regression and the *caret* package. Here, we use this same technique but replace *lm* with *rpart* to predict a categorical rather than numeric outcome. Script 6.5 gives the statements and relevant output with Figure 6.9 showing the resultant decision tree.

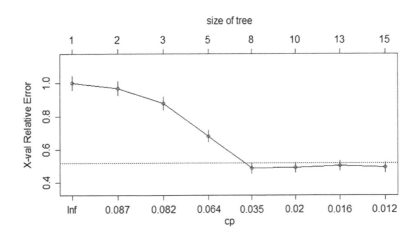

FIGURE 6.7   Decision tree complexity vs. relative error.

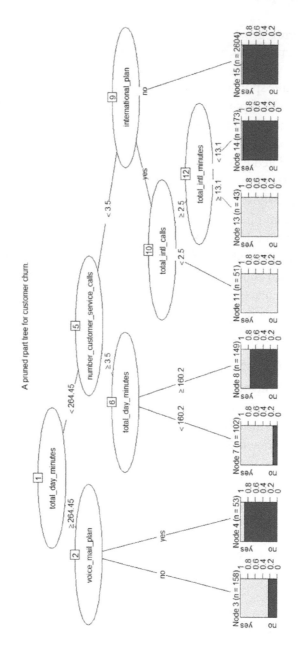

FIGURE 6.8 Top-level nodes of a pruned decision tree predicting customer churn.

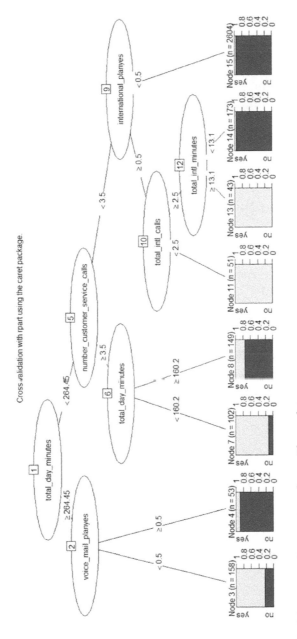

FIGURE 6.9 Decision tree using caret package: Churn data.

Script 6.5 Cross Validation with rpart: Churn Data

```
> # Build and test via a 10-fold cross validation a decision
> # tree using the caret package and the rpart function.

> #PREPROCESSING
> library(caret)
> library(rpart.plot)
> library(rpart)
> library(partykit)
> cv.control <- trainControl(method = "cv",number = 10)
> rpart.cv <- train(churn ~ ., data = churnTrain, method =
+                   "rpart", tuneLength=5, trControl = cv.control)
> rpart.cv

3333 samples
  19 predictor
   2 classes: 'yes', 'no'

Resampling: Cross-Validated (10 fold)
Summary of sample sizes: 2999, 2999, 3000, 3000, 3000, 3000, ...
Resampling results across tuning parameters:

  cp          Accuracy   Kappa
  0.01863354  0.9252981  0.6625721
  0.05279503  0.9171845  0.6053477
  0.07867495  0.8745851  0.3330197
  0.08488613  0.8676871  0.2518991
  0.08902692  0.8628859  0.2146019

Accuracy was used to select the optimal model using the largest
value.
The final value used for the model was cp = 0.01863354.

> # EVALUATE THE MODEL
> pred.cv <-  predict(rpart.cv, churnTest)
> conf.matrix <- table(churnTest$churn,pred.cv,dnn=c("Actual",
+ "Predicted"))
> conf.matrix

        Predicted
Actual  yes   no
   yes  141   83
    no   28 1415

Accuracy = 93.34 %
```

```
> # OUTPUT THE BEST TREE
> best.tree <- rpart.cv$finalModel
> plot(as.party(best.tree),
+ "Decision tree: Cross Validation with Caret Package Churn Data")
```

The script provides two items of particular interest. First, the tenfold cross validation (*cv*) is performed on the training data. With a training set size of 3333, we have 10 training samples each with 2999 or 3000 instances. An alternative to a single 10-fold cross validation is a repeated cross validation. To incorporate this, we use method = "repeatedcv" rather than "cv" and include the variable *repeats* to specify the number of repeated cross validations to perform.

A second item of primary interest deals with parameter tuning. There are two ways to tune algorithm parameters in the *caret package*. Parameter tuning can be done automatically with *tuneLength* or manually by specifying possible values for each parameter in a grid. The manual technique is interesting but requires more work. Here we use the *tunelength* variable to specify the total number of values to try for each algorithm parameter. With *tuneLength* set at 5 we see five entries in the complexity table. Each table value represents the outcome of one 10-fold cross validation. The cp value selected is based on model accuracy and gives a test set correctness score over 93% with 37% of the churners misclassified. Although not ideal, this outcome compares with our previous results with these data.

Here we barely touched on the many features of the *caret package*. It is well worth your time to take a closer look at the wealth of additional features this package has to offer. Next on our list is RWeka's J48 decision tree function!

6.4 BUILDING DECISION TREES: J48

In Chapter 5, we introduced the RWeka package by using the *GainRatioAttributeEval* function to help with attribute selection. Here we use *J48*, RWeka's java implementation of C4.5. Before we look at an example, be sure you have the RWeka package (CRAN library) installed on your machine. Also, RWeka contains several functions (also called packages) that have not been installed but are available for installation. To use all of the features offered by the RWeka package, you will eventually want to create an environment variable titled WEKA_HOME. You can assign this variable any valid path/folder name. As *J48* is installed in the base RWeka package, the environment variable assignment can wait.

For our experiment, we applied the J48 decision tree algorithm to the churn data. Script 6.6 lists the statements and relevant output for our experiment. Through experimentation, we determined 25 to be a good choice for the minimum number of instances per leaf (M). *WOW(J48)* lists all control options available for *J48*.

The test set accuracy of 94.12% compares well with our previous models for this data. The 68 churner misclassifications is also a positive result. However, the size of the tree (not displayed here) is less than optimal. Lastly, if you prefer to use a cross validation for model testing, you can do so with the function *evaluate_Weka_classifier*. Although not shown here, the cross validation is included in your copy of Script 6.6.

After experimenting with C5.0, *rpart*, and *J48*, you are well on your way to developing a firm base of machine learning knowledge about decision trees. No time to rest, we've just gotten started. Let's see what multiple model approaches have to offer!

Script 6.6 Train/Test Scenario with J48: Churn Data

```
> #PREPROCESSING
> library(RWeka)
> library(partykit)
> set.seed(100)
> # CREATE THE DECISION TREE
> churn.J48 <-J48(churn ~ ., data=churnTrain,
+                    control=Weka_control(M=25))
> churn.J48


J48 pruned tree
------------------


total_day_minutes <= 264.4
|    number_customer_service_calls <= 3
|    |    international_plan = no
|    |    |    total_day_minutes <= 223.2: no (2221.0/60.0)
|    |    |    total_day_minutes > 223.2
|    |    |    |    total_eve_minutes <= 242.3: no (296.0/22.0)
|    |    |    |    total_eve_minutes > 242.3
|    |    |    |    |    total_eve_minutes <= 267
|    |    |    |    |    |    total_day_minutes <= 239.5: no (25.0/3.0)
|    |    |    |    |    |    total_day_minutes > 239.5: yes (25.0/10.0)
|    |    |    |    |    total_eve_minutes > 267: yes (37.0/9.0)
|    |    international_plan = yes
.......


Number of Leaves  :      15
Size of the tree :       29


=== Confusion Matrix ===

> churn.pred <- predict(churn.J48, churnTest,type="class")
> churn.conf <- table(churnTest$churn,churn.pred,
+                    dnn=c("Actual","Predicted"))
> churn.conf
        Predicted
Actual  yes    no
```

```
  yes   156    68
  no     30  1413
```

```
> confusionP(churn.conf)
```

```
Accuracy = 94.12 %
> # PLOT THE MODEL
> plot(as.party(churn.J48),
+       "A J48 decision tree for customer churn")
```

6.5 ENSEMBLE TECHNIQUES FOR IMPROVING PERFORMANCE

When we make decisions, we usually rely on the help of others. In the same way, it makes intuitive sense to formulate classification decisions by combining the output of several machine learning models. Unfortunately, attempts at employing a multiple-model approach to classification have been met with mixed results. This can be explained in part by the fact that most models misclassify the same instances. However, some success has been seen with a multiple-model approach where each model is built by applying the same machine learning algorithm. In this section, we describe three techniques that make use of the same machine learning algorithm to construct multiple-model (ensemble) classifiers.

The approaches described here work best with unstable machine learning algorithms. Unstable algorithms show significant changes in model structure when slight changes are made to the training data. Many standard machine learning algorithms including decision trees have been shown to display this characteristic.

6.5.1 Bagging

Bagging (bootstrap aggregation) is a supervised learning approach developed by Leo Breiman (1996) that allows several models to have an equal vote in the classification of new instances. The same mining tool is most often employed to create the multiple models to be used. The models vary in that different training instances are selected from the same pool to build each model. Here's how the approach works:

1. Randomly sample several training datasets of equal size from the domain of training instances. Instances are sampled with replacement. This allows each instance to be present in several training sets.

2. Apply the machine learning algorithm to build a classification model for each training dataset. N sets of training data result in N classification models.

3. To classify unknown instance I, present I to each classifier. Each classifier is allowed one vote. The instance is placed in the class achieving the most votes.

Bagging can also be applied to estimation problems where predictions are numeric values. To determine the value of an unknown instance, the estimated output is given as the average of all individual classifier estimations.

6.5.2 Boosting

Boosting, introduced by Freund and Schapire (1996), is more complex. Boosting is like bagging in that several models are used to vote in the classification of new instances. However, there are two main differences between bagging and boosting. Specifically,

- Each new model is built based upon the results of previous models. The latest model concentrates on correctly classifying those instances incorrectly classified by prior models. This is usually accomplished by assigning weights to individual instances. At the start of training, all instances are assigned the same weight. After the latest model is built, those instances classified correctly by the model have their weights decreased. Instances incorrectly classified have their weights increased.

- Once all models have been built, each model is assigned a weight based on its performance on the training data. Because of this, better-performing models are allowed to contribute more during the classification of unknown instances.

As you can see, boosting builds models that complement one another in their ability to classify the training data. An obvious problem exists if the machine learning algorithm cannot classify weighted training data. The following is one way to overcome this problem:

1. Assign equal weights to all training data and build the first classification model.

2. Increase the weights of incorrectly classified instances and have the weights of correctly classified instances decreased.

3. Create a new training dataset by sampling instances with replacement from the previous training data. Select instances having higher weight values more frequently than instances with lower weight values. In this way, even though the machine learning algorithm cannot use instance weights per se, the most difficult to classify instances will receive increased exposure to subsequent models.

4. Repeat the previous step for each new model.

As incorrectly classified instances are sampled more frequently, the weight values still play their part in the model building process. However, the learning algorithm does not actually use the weight values during the model building procedure.

6.5.3 Boosting: An Example with C5.0

The C5.0 *trials* parameter with settings greater than 1 implements the boosting strategy described above. Let's give boosting a try! Return to Script 6.2, Set *trials =10* and execute the script.

Here is the relevant output:

```
Evaluation on training data (3333 cases):

Trial      Decision Tree
-----      ----------------
      Size       Errors

   0     8   237( 7.1%)
   1     4   442(13.3%)
   2     5   469(14.1%)
   3     4   463(13.9%)
   4    12   923(27.7%)
   5     7   536(16.1%)
   6     7   355(10.7%)
   7     7   472(14.2%)
   8     4   386(11.6%)
   9     8   340(10.2%)
boost         197( 5.9%)   <<

> #Test Set Confusion Matrix

       Predicted
Actual  yes    no
   yes  119   105
   no     6  1437

Accuracy = 93.34 %
```

The output shows the accuracy of each of the ten trees when individually presented with the training data. Trial 0 (step 1 of the algorithm) gives the tree obtained without boosting. The Errors column shows the error obtained by applying each new tree to the training data. Notice that the Trial 0 tree outperforms each subsequently created tree. The line labeled *boost* shows the training set accuracy when the models work in unison, each getting one vote. The boosted model does slightly outperform the Trial 0 model on the training data. However, when the boosted model is applied to the test data, we see that the number of incorrectly classified churners has increased from 83 to 105.

This simple experiment supports what is known about the multiple model approach in that it doesn't necessarily improve performance. Obviously, more experimentation with the *trials* variable may give us a better result. As a general rule, boosting does best in environments where noisy data is rare. The next technique overcomes some of the problems seen with both bagging and boosting.

6.5.4 Random Forests

The random forest approach (Breiman, 2001) is similar to bagging in that it employs a collection of decision trees and a majority rules approach. However, unlike bagging where the same attributes are used to build each tree, a random forest selects random subsets of attributes to build the individual trees. Here is the approach:

1. Choose the number N of decision trees for the forest.

2. Repeat the following N times:

   a. Randomly select a subset of instances with replacement from the set of training data.

   b. Randomly select a subset of attributes from the set of chosen attributes.

   c. Build a decision tree using the selected attributes and instances.

To illustrate the approach, let's return to the *creditScreening* dataset described in Chapter 5 where we used *Bayes* classifier with this goal:

Build a predictive model using the instances of the credit card screening dataset able to predict which future credit card applicants to accept and which to reject. Benchmarks for predictive accuracy with previously unseen data include an overall model correctness of 85%, a 10% maximum for incorrect denials, and a 5% maximum for incorrect accepts.

Script 6.7 lists the statements and edited output for our experiment that applies the *randomForest* function to the *creditScreening* dataset. The default settings for the number of trees is 500. The default for the number of attributes is the truncated square root of the number of input attributes. With 15 input attributes, the default is 3.

A point of interest in the call to *randomForest* is the setting of *na.action* to *na.roughfix*. With this setting, a missing attribute value is replaced by the median for the attribute. If the attribute is categorical, the missing value is replaced with the attribute mode. If more than one mode exists, one of the modes is chosen at random. It is important to note that test set instances with one or more missing items are not classified.

A second point of interest is the importance option. Importance is a method of the randomForest object that, when given a model, returns a list of the input attributes together with a heuristic measure of their importance. This is possible as each tree of a random forest uses a randomized subset of the total attribute space. The attributes of the trees within the forest that perform best obtain the highest importance scores.

Notice there is some disagreement between the best attributes as determined by RWeka's *GainRatioAttributeEval* in Script 5.11 and the importance ranking shown here. However, both approaches show attribute nine at the top of the list. You can learn more about the two types of importance measures (type 1 & type 2) with help(importance).

Step 2a of the random forest algorithm tells us that each tree is built with a subset of the entire set of training data. Once a tree is built, the unused instances are given to the tree for classification. The OOB (out-of-bag) estimate of error rate of 14.78% is the classification error rate for these unused instances. The confusion matrix obtained in classifying the OOB instances shows a classification correctness of about 85%. OOB is particularly useful when test data is lacking.

The test set error rate of 91.52% is satisfactory. The false positive rate of 0.107 is an improvement over the results seen with Bayes' classifier (Script 5.11) but still misses the 5% mark set in our goal statement.

Script 6.7: Random Forest Application: Creditscreening Data

```
> library(randomForest)


> #PREPROCESSING
> set.seed(1000)
> # Randomize and split the data for 2/3 training, 1/3 testing
> credit.data <- creditScreening
> index <- sample(1:nrow(credit.data), 2/3*nrow(credit.data))
> credit.train <- credit.data[index,]
> credit.test <-  credit.data[-index,]


> # CREATE THE FOREST

> credit.forest <- randomForest(class ~ ., data=credit.train,
+                               na.action = na.roughfix,
+                               importance=TRUE)
> credit.forest

Call:
 randomForest(formula = class ~ ., data = credit.train, importance
= TRUE,      na.action = na.roughfix)
               Type of random forest: classification
                     Number of trees: 500
No. of variables tried at each split: 3

        OOB estimate of error rate: 14.78%


Confusion matrix:
      -   + class.error
- 219  37   0.1445312
+  31 173   0.1519608
> # List the most important attributes.
```

```
> importance(credit.forest,type=2)
          MeanDecreaseGini     MeanDecreaseGini
one              2.509096 two  15.014976
three           15.980267 four      3.296809
five             3.178326 six      26.621870
seven            8.667658 eight    20.025745
nine            59.669127 ten      10.152847
eleven          19.931858 twelve    2.277566
thirteen         2.193586 fourteen 13.809834
fifteen         19.308132

> credit.pred <-predict(credit.forest, credit.test)
> credit.perf<- table(credit.test$class, credit.pred,
+ dnn=c("actual", "Predicted"))
> credit.perf
       Predicted
actual   -    +
     - 109   13
     +   6   96
> confusionP(credit.perf)
  Correct= 205 Incorrect= 19
  Accuracy = 91.52 %
```

Random forests overcome several of the shortcomings seen with decision trees one of which is the ability to classify in domains where the number of attributes greatly exceeds the number of instances. Also, as the trees in the forest are built with different sets of attributes, it is possible to determine those attributes best able to consistently build accurate tree structures.

On the negative side, random forests can be computationally expensive and can require large amounts of memory. Also, with the large number of trees, the production rules representing the forest are at best difficult to understand. We will take a closer look at the rules generated by a random forest in Chapter 7. Regression trees are next!

6.6 REGRESSION TREES

Regression trees represent an alternative to statistical regression. Regression trees get their name from the fact that most statisticians refer to any model that predicts numeric output as a regression model. Essentially, regression trees take the form of decision trees where the leaf nodes of the tree are numeric rather than categorical values. The value at an individual leaf node is computed by taking the numeric average of the output attribute for all instances passing through the tree to the leaf node position.

Regression trees are more accurate than linear regression equations when the data to be modeled is nonlinear. However, regression trees can become quite cumbersome and difficult to interpret. For this reason, regression trees are sometimes combined with linear regression to form what are known as *model trees*. With model trees, each leaf node

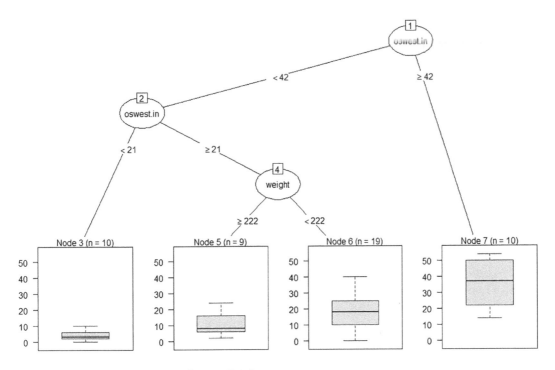

FIGURE 6.10   A regression tree for OFCM data.

of a partial regression tree represents a linear regression equation rather than an average of attribute values. By combining linear regression with regression trees, the regression tree structure can be simplified in that fewer tree levels are necessary to achieve accurate results.

Figure 6.10 shows a regression tree for the OFCM dataset introduced in Script 5.5. Script 6.8 displays the modifications to Script 5.5 with *rpart* replacing *lm*. As before, the goal is to develop a predictive model able to determine end of treatment pain level. The statements defining the regression tree offer at least two points of interest. First, patients with a before treatment pain level (*oswest.in*) greater than 42 are likely to see a less than optimal outcome. Second, patient weight might play a role in treatment success. Also, the residual standard error of 10.324 is a 15% improvement over that seen with the linear regression model. As this dataset is small, these conclusions are best taken as hypotheses to be tested with further research.

Script 6.8 Regression Tree for OFC Data

```
> # Our goal is to build a regression tree to
> # predict the after treatment oswestry value. Only
> # values known at the beginning of treatment are
> # used for input.

> # PREPROCESSING
```

```
> library(rpart)
> library(partykit)
> # Remove gender & attributes measured at the end of treatment
>   ofc.data <- OFCM[-c(1,6,8,10)]
> # Check for correlations
> # round(cor(ofc.data),3)
> set.seed(100)
> osw.rpart <- rpart(oswest.out ~ ., data =ofc.data,na.action=na.omit)
> osw.rpart

n= 48
node), split, n, deviance, yval
      * denotes terminal node

 1) root 48 10623.670 17.58333
   2) oswest.in< 42 38   4285.053 12.84211
     4) oswest.in< 21 10     91.600  3.80000 *
     5) oswest.in>=21 28   3083.857 16.07143
      10) weight>=222 9    456.000 11.33333 *
      11) weight< 222 19   2330.105 18.31579 *
   3) oswest.in>=42 10   2238.400 35.60000 *

> #summary(osw.rpart)
> plot(as.party(osw.rpart))
> output <- (mmr.stats(osw.rpart,ofc.data,ofc.data$oswest.out))

[1] "Mean Absolute Error ="
[1] 8.0738
[1] "Mean Squared Error ="
[1] 106.5855
[1] "Root Mean Squared Error="
[1] 10.324
[1] "Residual standard error="
[1] 10.324
```

6.7 CHAPTER SUMMARY

Decision trees have several advantages.

- Decision trees are easy to understand and map nicely to a set of production rules.

- Decision trees have been successfully applied to real problems.

- Decision trees make no prior assumptions about the nature of the data.

- Decision trees are able to build models with datasets containing numerical as well as categorical data.

As with all machine learning algorithms, there are several issues surrounding decision tree usage. Specifically,

- Output attributes must be categorical, and multiple output attributes are not allowed.

- Decision tree algorithms are *unstable* in that slight variations in the training data can result in different attribute selections at each choice point within the tree. The effect can be significant as attribute choices affect all descendent subtrees.

- Trees created from numeric datasets can be quite complex as attribute splits for numeric data are typically binary.

One decision tree model we didn't cover here is the conditional inference tree where splits are based on attribute significance as determined through permutation testing (Chapter 10). The *ctree( )* function found in the party package is a conditional inference tree implementation available with R's Rattle package (Chapter 7).

Multiple-model methods such as bagging, boosting, and random forests can sometimes improve model performance. These approaches work best with unstable machine learning algorithms. Regression trees and model trees provide additional options for building models having numeric output.

6.8 KEY TERMS

- *Bagging.* A supervised learning approach that allows several models to have an equal vote in the classification of new instances.

- *Boosting.* A supervised learning approach that allows several models to take part in the classification of new instances. Each model has an associated weight that is applied toward new instance classification.

- *Model Tree.* A partial regression tree where each leaf node of the tree represents a linear regression equation rather than an average of attribute values.

- *Random Forest.* A majority rules approach where a collection of decision trees is built with random subsets of attributes. Each tree has an equal vote in the classification of new instances.

- *Regression Tree.* A decision tree where the leaf nodes of the tree are numeric rather than categorical values.

EXERCISES

Review Questions

1. Describe the two types of importance available for *randomForest* models.

2. How do regression trees and model trees differ? How are they the same?

3. What are the key differences between boosting and the random forest approach?

4. The churn dataset was used in conjunction with C5.0, *rpart* with a training/test set scenario, *rpart* with cross validation, and *J48*. Summarize the test set results for each of these experiments.

Experimenting with R

1. Use Script 6.2 to experiment with the *trials* parameter by giving it values between 10 and 100 in increments of 10. Make a table of your results that for each trial shows overall accuracy and the total number of misclassified churners.

2. Consider Script 6.2. Set the trials parameter at the value giving the fewest misclassified churners in Exercise 1. Starting with *minCases=10*, experiment with alternative settings for *minCases* in increments of 10 up to and including 100. Make a table to show overall accuracy, and the total number of misclassified churners within the created tree for each value of *minCases*. Given that your goal is to minimize the misclassification of churners, make a statement about a best choice for *minCases*.

3. Load Script 6.7 and experiment with the *randomForest* function settings for the number of trees and the number of attributes using the *ntree* and *mtry* parameters. Try at least five alternate settings for each parameter. Summarize your results. Are you able to improve model test set performance?

4. Perform a cross validation with *rpart* and the *caret package* using the *cardiology-Mixed* dataset. The goal is to identify individuals who have had at least one heart attack. Select a random subset of 200 instances for building the model. The remaining 103 instances are used for testing your model. Vary the *tuneLength* parameter using at least five different values. Summarize your results.

5. Use the *cardiologyMixed* dataset together with *rpart* and a training/test set scenario similar to Script 6.4 to build a model for determining those patients who have had a heart attack. Improve your model by using the complexity table to prune the created decision tree. Provide the test set confusion matrix for both the unpruned and pruned model.

6. Use Script 6.25: found in the scripts.zip file for Chapter 6 to identify the first 10 instances in the ordered list that are incorrectly predicted not to churn. Next, modify the script so it uses *rpart* rather than C5.0 to build the predictive model. In the same manner, output the predictions as probabilities. How many of the 10 instances incorrectly classified by C5.0 are incorrectly classified by *rpart*? What do you conclude?

Programming Project—Missing Data

The creditScreening data frame (creditScreening.csv) contains several instances with missing data. Missing values of numeric or integer type show as an NA. Missing categorical attribute values are simply blank. Your job is to replace missing numeric or integer items using *rpart*. Here is what to do.

1. Create the function saveNAS by modifying the removeNAS function so when given a data frame, it returns a data frame containing all instances having at least one missing numeric attribute value.

2. Make calls to both functions as follows:

   a. goodDF <- removeNAS(creditScreening)

   b. badDF <- saveNAS(creditScreening)

3. Examine the contents of badDF. You will notice that attributes *two* and *fourteen* are the only numeric/integer type attributes with missing values. Write a script that builds a regression tree with *rpart* where the output attribute is designated as *two*.

4. Use *predict* to apply your model to badDF. *Predict* will estimate values for attribute *two* within badDF. Replace all missing values for attribute *two* with predicted values.

5. Repeat steps 3 and 4 but replace attribute *two* with attribute *fourteen*.

6. Combine goodDF and badDF to form a new data frame titled *creditScreeningNew*.

Installed Packages and Functions

Package Name	Function(s)
base/ stats	*data, library, na.action, na.omit, nrow, plot, predict round, set.seed, sample, summary,table,*
C50	*C5.0*
Caret	*train, trainControl*
party.kit	*as.party*
randomForest	*randomForest*
RWeka	*J48, GainRatioAttributeEval*
rpart	*rpart*
rpart.plot	*prp, rpart.plot, rpart.rules*

## CHAPTER 7

Rule-Based Techniques

In This Chapter

- Decision Tree Rules

- A Covering Rule Algorithm

- Association Rules

- The Rattle Package

IN THIS CHAPTER, WE continue our discussion of supervised learning with a focus on rule-based machine learning techniques. The emphasis of Section 7.1 is on decision tree rules. In Section 7.2, we outline a fundamental covering rule algorithm. We then apply RWeka's *JRip* covering algorithm to the churn data introduced in Chapter 6. In Section 7.3, we demonstrate an efficient technique for generating association rules. The *Apriori(RWeka)* association rule function is then utilized to find interesting relationships in a customer database of grocery store purchases. The focus of Section 7.4 is on Rattle, a graphical user interface (GUI) supporting many of the preprocessing, modeling and evaluation methods discussed throughout your text. We use Rattle's interface to generate production rules with *rpart*, model customer churn with the *randomForest* function, and generate association rules with the *apriori(arules)* function.

7.1 FROM TREES TO RULES

In Chapter 1, you saw how a decision tree can be mapped to a set of production rules by writing one rule for each path of the tree. As rules tend to be more appealing than trees, several variations of the basic tree to rule mapping have been studied. Most variations focus on simplifying and/or eliminating existing rules. To illustrate the rule simplification process, consider the decision tree in Figure 6.6. A rule created by following one path of the tree is shown here:

IF Age < 44 & Gender = Male & Credit Card Insurance = No

THEN Life Insurance Promotion = No

The antecedent conditions for this rule covers 4 of the 15 instances with a 75% accuracy. Let's simplify the rule by eliminating the antecedent condition for *age*. The simplified rule takes the form:

IF Gender = Male & Credit Card Insurance = No

THEN Life Insurance Promotion = No

By examining Table 6.1, we see the antecedent of the simplified rule covers six instances. As the rule consequent covers five of the six instances, the accuracy of the simplified rule is approximately 83.3%. Therefore, the simplified rule is more general and accurate than the original rule! At first thought, it seems hard to believe that removing a conditional test can actually improve the accuracy of a rule. However, closer examination shows why eliminating the test gives a better result. To see this, notice that removing the *age* attribute from the rule is equivalent to deleting the attribute from the tree in Figure 6.6. In doing so, the three instances following the path *age >= 44* must now follow the same path as those instances traversing *age <44*. All three instances with *age >= 44* are members of the *life insurance promotion = no* class. Two of these three instances are of male gender with *credit card insurance = no*. Both instances satisfy the preconditions and consequent condition of the simplified rule. Because of this, the preconditions for the new rule are satisfied by six instances, five of which have *no* as the value of *life insurance promotion*.

Most decision tree implementations automate the process of rule creation and simplification. Once rules have been simplified and/or eliminated, the rules may be ordered so as to minimize error. Finally, a default rule is chosen. The default rule states the classification of an instance not meeting the preconditions of any listed rule. Let's use the *Spam email* dataset to take a look at how C5.0 generates decision tree rules.

7.1.1 The Spam Email Dataset

The Spam email dataset contains 4601 instances, 2788 of which are valid email messages. The remaining instances are classified as spam. All 57 input attributes are continuous with no missing items. The output attribute is categorical with 0 representing a valid email message. A complete description about the origins and nature of the data is given in the *description* sheet of Spam.xlsx. This file together with Spam.csv is found in your supplementary materials.

Let's build a rule-based model using C5.0 that can distinguish between valid and invalid email messages. Our goal is to use the model to determine whether a new email message will be delivered to a user's inbox or placed in the junk folder. Desirable model characteristics include fast processing and high accuracy levels where the majority of errors classify spam email as valid. That is, we would rather have the user delete spam from the inbox than miss valid messages that reside in their junk folder!

7.1.2 Spam Email Classification: C5.0

To get started, import *Spam.csv* into RStudio, and load Script 7.1 into your RStudio editor. The statements and relevant output are displayed in *Script 7.1 Analyzing the Spam dataset.* Several statements have been commented out to conserve space. Removing the comment symbols before executing the script will give you a better overall picture of the data.

The first statement of interest is the call to C5.0 with *rules=TRUE*. In this case, the output of C5.0 is a set of rules rather than a decision tree. The first five rules are listed in Script 7.1. The ordering of the rules is arbitrary. If none of the rules apply, the instance is given the classification of the most frequent class within the data. This is seen here as *Default class: 0.* The number of rules generated can be controlled by varying the setting for *minCases.* Larger values for *minCases* will also reduce overall classification accuracy.

To better understand rule accuracy, consider the fifth rule stated as:

```
Rule 5: (616/8, lift 1.6)
        Remove <= 0.07
        Hp > 0.39
        -> class 0 [0.985]
```

This rule tells us that 616 instances satisfy the antecedent condition of the rule. Of the 616 instances, 8 belong to class 1. This gives us a rule accuracy of 0.985 computed using the Laplace ratio $(n - m + 1)/(n + 2)$ where $n = 616$ and $m = 8$. Lift is determined by dividing rule accuracy (0.985) by the relative frequency of the class within the training data (0.6025). Rules with larger lift values are particularly useful when dealing with imbalanced data where classification accuracy is often a poor metric.

Script 7.1 Analyzing the Spam Dataset

```
> library(C50)
> library(partykit)


> # PREPROCESSING


> # Randomize and split the data for 2/3 training, 1/3 testing
> set.seed(100)
> spam.data <- Spam
> # summary(spam.data)
> # removeNAS(spam.data)
> index <- sample(1:nrow(spam.data), 2/3*nrow(spam.data))
> spam.train <- spam.data[index,]
> spam.test <-  spam.data[-index,]


> # BUILD THE MODEL
> Spam.C50 <- C5.0(Spam ~ ., data = spam.train,
+                  control = C5.0Control(minCases=2),rules=TRUE)
> summary(Spam.C50)
```

```
Call:
C5.0.formula(formula = Spam ~ ., data = spam.train, control =
C5.0Control(minCases = 2), rules = TRUE)

Rules:

Rule 1: (424, lift 1.7)  Rule 2: (434, lift 1.7)
     X0 <= 0.25                George > 0.08
     George > 0.38             Exclamation <=0.142
     Exclamation <= 0.375      -> class 0 [0.998]
     ->  class 0  [0.998]


Rule 3: (95, lift 1.6)        Rule 4: (103, lift 1.6)
     Cs > 0.08                 Meeting > 1.1
     ->  class 0  [0.990]      -> class 0 [0.990]


Rule 5: (616/8, lift 1.6)
     Remove <= 0.07
     Hp > 0.39
     ->  class 0  [0.985]


     ->  class 0  [0.984]
Default class: 0

     Attribute usage:

        96.58%      Dollar.Sign
        55.62%      Exclamation
        54.87%      Remove
        50.93%      Hp
        49.23%      George
        47.86%      Capital.Run.Length.Longest
        .........
> #TEST THE MODEL

> pred.C50 <-  predict(Spam.C50, spam.test)
> C50.perf<- table(spam.test$Spam, pred.C50, dnn=c("actual",
+ "Predicted"))
> C50.perf

      Predicted
actual   0   1
     0 894  46
     1  71 523

> confusionP(C50.perf)
  Correct= 1417 Incorrect= 117
  Accuracy = 92.37 %
```

Attributes listed at the top of the (partial) *attribute usage* list are most relevant for predicting outcome. The 92% overall accuracy is good but almost 5% of the valid emails were incorrectly classified as junk mail. To lower the number of incorrect valid email classifications, let's look at prediction probability values. A return to Script 5.8 helps set up the experiment displayed in Script 7.2.

As with Script 5.8, the *ifelse* statement gives us an opportunity to alter the minimal probability value for classifying an email as spam. If we use 0.50, our results will be identical to those in Script 7.1. Here we change the cutoff requirement from 0.50 to 0.75. The confusion matrix shows that overall classification accuracy is maintained while lowering the 5% mark for incorrect valid email classifications to less than 3%. The price paid is an increase in the amount of spam making it to the inbox. We are definitely on the right track. An opportunity for additional work with the Spam dataset is offered in the end-of-chapter exercises. The next section offers an approach to rule generation that does not rely on a decision tree structure.

Script 7.2 Experimenting with Probabilities: Spam Data

```
# DETERMINE PROBABILITIES OF CLASS MEMBERSHIP
> pred.C50 <-  predict(Spam.C50, spam.test,type = "prob")
> pred.C50 <- cbind(spam.test$Spam,pred.C50)
> pred.df  <- data.frame(pred.C50)
> colnames(pred.df)<- c("spam","no","yes")

> # EXPERIMENT & CREATE CONFUSION MATRIX
> spam.results <- ifelse(pred.df$yes > 0.75,2,1) # > spam
> spam.pred <- factor(spam.results,labels=c("no","yes"))
> my.conf <- table(spam.test$Spam,spam.pred,dnn=c("Actual",
+ "Predicted"))
> my.conf

      Predicted
Actual  no yes
     0 913  27
     1 110 484

> confusionP(my.conf)
  Correct= 1397 Incorrect= 137
  Accuracy = 91.07 %
```

7.2 A BASIC COVERING RULE ALGORITHM

A popular class of rule generators uses what is referred to as a *covering* approach. Instances are covered by a rule if they satisfy the rule's preconditions. For each class, the goal is to create a set of rules that maximizes the total number of covered within-class instances while at the same time minimizing the number of covered nonclass instances.

Here we outline a straightforward covering technique based on the PRISM algorithm (Cendrowska, 1987). The method starts with a single rule whose antecedent condition is empty and whose consequent is given as the class representing the instances to be covered. At this point, the domain of all dataset instances is covered by the rule. That is, when the rule is executed, it classifies every instance as a member of the class specified in the rule consequent. In order to restrict rule coverage to those instances within the class, input attribute values are considered one by one as possible additions to the rule antecedent. For each iteration, the attribute-value combination chosen is the one that best represents those instances within the class. A rule is added to the list of covering rules once the preconditions of the rule are satisfied solely by members of the class in question or the attribute list has been exhausted. After the rule has been added, the instances covered by the rule are removed from the pool of uncovered instances. The process continues until all instances from the class in question are covered by one or more rules. The method is then repeated for all remaining classes. Several variations of this general technique exist but the covering concept is the same.

In practice, the method given here is but the first step in constructing a set of useful rules. This is the case as rules based solely on a single dataset, although accurate for the instances within the dataset, may not do well when applied to the larger population. Techniques for generating useful covering rules are often based on the concept of *incremental-reduced-error pruning*. The general idea is to divide the instances using a 2:1 ratio into two sets. The larger (grow) set is used during the growing phase of the algorithm for growing perfect rules. The smaller (prune) set is applied during the pruning phase for removing one or more antecedent conditions from perfect rules.

During the growing phase, the *grow* set is used for creating a perfect rule R for a given class C. Next, a heuristic measure of R's worth such as classification correctness is computed by applying R to the *prune* set. After this, rule antecedent conditions are removed from R one by one each time creating a new rule R- which contains one less precondition than the previous rule. Upon each removal, a new worth score for R- is computed. This process continues as long as the worth of each new rule is greater than the worth of the original perfect rule. Upon termination of this process, the rule with the highest worth is added to the list of rules for C, and the instances covered by this best rule are removed from the instance set. This procedure repeats for C as long as at least one instance remains in both grow and prune. This entire process repeats itself for all remaining classes within the data. An enhanced version of this basic technique that includes an optimization phase is known as the RIPPER (*repeated incremental pruning to produce error reduction*) algorithm (Cohen, 1995) a variation of which is implemented in the *JRip* function found within the RWeka package. Let's see how *JRip* does with the *churn* dataset.

7.2.1 Generating Covering Rules with JRip

Script 7.3 displays the statements and edited output for our covering rules experiment. Preprocessing includes a check for missing items in both the training and test data. The call to *JRip* creates 12 covering rules after automatically setting the randomization seed at 1. It is important to note that the rules must be executed in the order given!

Script 7.3 lists the first four rules and the last rule. The first rule correctly identifies 72 of 73 training set instances. The second and third rules taken together correctly classify 126 instances as churners. The fourth rule incorrectly classifies 8 of 57 instances. The final rule is the last of the 12 rules. The first 11 rules are posed to identify churners. The last rule covers all instances not covered by the first 11 rules. The rule misclassifies 87 of the remaining 2920 instances and simply states that if an instance does not satisfy any one of rules 1 through 11, then the instance is classified as a non-churner.

The summary statement contains a confusion matrix based on the training data. Of the 483 churners, we see 87 incorrectly classified as non-churners. This represents over 18% of the churning population found within the training data. This percent increases to over 27% when the rules are applied to the test data. As our main concern is the identification of candidate churners, the 95.74 % overall test set correctness score is of little consequence. Experimenting with *JRip*'s parameters as well as an investigation of outlier instances are two avenues for possible improvement. In Section 7.4, we will find out if a random forest approach gives a better result.

Script 7.3 Generating Covering Rules with JRip

```
> #PREPROCESSING
> library(RWeka)

> # Check for NA's but do not delete NA instances.
> x<- removeNAS(churnTrain)
[1] "number deleted"
[1] 0
> Y <-removeNAS(churnTest)
[1] "number deleted"
[1] 0

> # CREATE THE RULE MODEL
> churn.Rip <-JRip(churn ~ ., data=churnTrain)
> churn.Rip

JRIP rules:
===========

(total_eve_minutes >= 214.3) and (voice_mail_plan = no) and
(total_day_minutes >= 263.8) and (total_night_minutes >= 107.3) =>
churn=yes (73.0/1.0)

(international_plan = yes) and (total_intl_calls <= 2) =>
churn=yes (59.0/0.0)
```

```
(number_customer_service_calls >= 4) and (total_day_minutes <=
159.4) and (total_eve_minutes <= 231.3) and (total_night_minutes
<= 255.3) => churn=yes  (67.0/0.0)

(total_day_minutes >= 221.9) and (total_eve_minutes >= 241.9) and
(voice_mail_plan = no) and (total_night_minutes >= 173.2) =>
churn=yes  (57.0/8.0)
......

=> churn=no  (2920.0/87.0)

Number of Rules : 12

> summary(churn.Rip)

=== Summary ===

Correctly Classified Instances        3229              96.8797 %
Incorrectly Classified Instances       104               3.1203 %
Kappa statistic                         0.866
Mean absolute error                     0.0596
Root mean squared error                 0.1727
Relative absolute error                24.0407 %
Root relative squared error            49.0463 %
Total Number of Instances            3333

=== Confusion Matrix ===

    a    b    <-- classified as
  396   87 |     a = yes
   17 2833 |     b = no

> # TEST THE RULE MODEL
> churn.pred <- predict(churn.Rip, churnTest)
> churn.conf <- table(churnTest$churn,churn.pred,dnn=c("Actual",
+ "Predicted"))
> churn.conf

        Predicted
Actual   yes    no
   yes   163    61
    no    10  1433

> confusionP(churn.conf)
  Correct= 1596 Incorrect= 71
  Accuracy = 95.74 %
```

7.3 GENERATING ASSOCIATION RULES

Affinity analysis is the general process of determining which things go together. A typical application is market basket analysis, where the desire is to determine those items likely to be purchased by a customer during a shopping experience. The output of the market basket analysis is a set of associations about customer-purchase behavior. The associations are given in the form of a special set of rules known as association rules. The association rules are used to help determine appropriate product marketing strategies. In this section, we describe an efficient procedure for generating association rules.

7.3.1 Confidence and Support

Association rules are unlike traditional classification rules in that an attribute appearing as a precondition in one rule may appear in the consequent of a second rule. In addition, traditional classification rules usually limit the consequent of a rule to a single attribute. Association rule generators allow the consequent of a rule to contain one or several attribute values. To show this, suppose we wish to determine if there are any interesting relationships to be found in customer purchasing trends among the following four grocery store products:

- Milk
- Cheese
- Bread
- Eggs

Possible associations include the following:

1. If customers purchase milk, they also purchase bread.
2. If customers purchase bread, they also purchase milk.
3. If customers purchase milk and eggs, they also purchase cheese and bread.
4. If customers purchase milk, cheese, and eggs, they also purchase bread.

The first association tells us that a customer who purchases milk is also likely to purchase bread. The obvious question is "How likely will the event of a milk purchase lead to a bread purchase?" To answer this, each association rule has an associated confidence. For this rule, confidence is the conditional probability of a bread purchase given a milk purchase. Therefore, if a total of 10,000 customer transactions involve the purchase of milk, and 5000 of those same transactions also contain a bread purchase, the confidence of a bread purchase given a milk purchase is 5000/10,000 = 50%.

Now consider the second rule. Does this rule give us the same information as the first rule? The answer is no! With the first rule, the transaction domain consisted of all customers who had made a milk purchase. For this rule, the domain is the set of all customer transactions that show the purchase of a bread item. As an example, suppose we have a total of 20,000 customer transactions involving a bread purchase and of these, 5000 also involve a milk purchase. This gives us a confidence value for a milk purchase given a bread purchase of 25% vs. 50% for the first rule.

Although the third and fourth rules are more complex, the idea is the same. The confidence for the third rule tells us the likelihood of a purchase of both cheese and bread given a purchase of milk and eggs. The confidence for the fourth rule tells us the likelihood of a bread purchase given the purchase of milk, cheese, and eggs.

One important piece of information that a rule confidence value does not offer is the percent of all transactions containing the attribute values found in an association rule. This statistic is known as the *support* for a rule. Support is simply the minimum percentage of instances (transactions) in the database that contain all items listed in a specific association rule. In the next section, you will see how item sets use support to set limits on the total number of association rules for a given dataset.

7.3.2 Mining Association Rules: An Example

When several attributes are present, association rule generation becomes unreasonable because of the large number of possible conditions for the consequent of each rule. Special algorithms have been developed to generate association rules efficiently. One such algorithm is the *apriori* algorithm (Agrawal et al., 1993). This algorithm generates what are known as *item sets*. Item sets are attribute-value combinations that meet a specified coverage requirement. Those attribute-value combinations that do not meet the coverage requirement are discarded. This allows the rule generation process to be completed in a reasonable amount of time.

Apriori association rule generation using item sets is a two-step process. The first step is item set generation. The second step uses the generated item sets to create a set of association rules. We illustrate the idea with the subset of the credit card promotion database shown in Table 7.1. The *income range* and *age* attributes have been eliminated.

TABLE 7.1   A Subset of the Credit Card Promotion Database

Magazine Promotion	Watch Promotion	Life Insurance Promotion	Credit Card Insurance	Gender
Yes	No	No	No	Male
Yes	Yes	Yes	No	Female
No	No	No	No	Male
Yes	Yes	Yes	Yes	Male
Yes	No	Yes	No	Female
No	No	No	No	Female
Yes	No	Yes	Yes	Male
No	Yes	No	No	Male
Yes	No	No	No	Male
Yes	Yes	Yes	No	Female

To begin, we set the minimum attribute-value coverage requirement at four items. The first item set table created contains single-item sets. Single-item sets represent individual attribute-value combinations extracted from the original dataset. We first consider the attribute magazine promotion. Upon examining the table values for magazine promotion, we see seven instances have a value of yes and three instances contain the value no. The coverage for magazine promotion = yes exceeds the minimum coverage of four and, therefore, represents a valid item set to be added to the single-item set table. As magazine promotion = no with coverage of three does not meet the coverage requirement, it is not added to the single-item set table. Table 7.2 shows all single-item set values from Table 7.1 that meet the minimum coverage requirement.

We now combine single-item sets to create two-item sets with the same coverage restriction. We need only consider attribute-value combinations derived from the single-item set table. Let's start with *magazine promotion = yes* and *watch promotion = no*. Four instances satisfy this combination; therefore, this will be our first entry in the two-item set table (Table 7.3). We then consider *magazine promotion = yes* and *life insurance promotion = yes*. As there are five instance matches, we add this combination to the two-item set table. We now try *magazine promotion = yes* and *life insurance promotion = no*. As there are only two matches, this combination is not a valid two-item set entry. Continuing this process results in a total of 11 two-item set table entries.

TABLE 7.2 Single-Item Sets

Single-Item Sets	Number of Items
Magazine Promotion = Yes	7
Watch Promotion = Yes	4
Watch Promotion = No	6
Life Insurance Promotion = Yes	5
Life Insurance Promotion = No	5
Credit Card Insurance = No	8
Gender = Male	6
Gender = Female	4

TABLE 7.3 Two-Item Sets

Two-Item Sets	Number of Items
Magazine Promotion = Yes & Watch Promotion = No	4
Magazine Promotion = Yes & Life Insurance Promotion = Yes	5
Magazine Promotion = Yes & Credit Card Insurance = No	5
Magazine Promotion = Yes & Gender = Male	4
Watch Promotion = No & Life Insurance Promotion = No	4
Watch Promotion = No & Credit Card Insurance = No	5
Watch Promotion = No & Gender = Male	4
Life Insurance Promotion = No & Credit Card Insurance = No	5
Life Insurance Promotion = No & Gender = Male	4
Credit Card Insurance = No & Gender = Male	4
Credit Card Insurance = No & Gender = Female	4

The next step is to use the attribute-value combinations from the two-item set table to generate three-item sets. Reading from the top of the two-item set table, our first possibility is

Magazine Promotion = Yes & Watch Promotion = No & Life Insurance Promotion = Yes.

As only one instance satisfies the three values, we do not add this combination to the three-item set table. However, two three-item sets do satisfy the coverage criterion:

Watch Promotion = No & Life Insurance Promotion = No & Credit Card Insurance = No.

Life Insurance Promotion = No & Credit Card Insurance = No & Gender = Male.

As there are no additional member set possibilities, the process proceeds from generating item sets to creating association rules. The first step in rule creation is to specify a minimum rule confidence. Next, association rules are generated from the two- and three-item set tables. Finally, any rule not meeting the minimum confidence value is discarded.

Two possible two-item set rules are

IF *Magazine Promotion = Yes*

THEN *Life Insurance Promotion = Yes* (5/7)

IF *Life Insurance Promotion = Yes*

THEN *Magazine Promotion = Yes* (5/5)

The fractions at the right indicate the rule accuracy (confidence). For the first rule, there are five instances where magazine promotion and life insurance promotion are both yes. There are seven total instances where magazine promotion = yes. Therefore, in two situations, the rule will be in error when predicting a life insurance promotion value of yes when magazine promotion = yes. If our minimum confidence setting is 80%, this first rule will be eliminated from the final rule set. The second rule states that magazine promotion = yes any time life insurance promotion = yes. The rule confidence is 100%. Therefore, the rule becomes part of the final output of the association rule generator. Here are three of several possible three-item set rules:

IF *Watch Promotion = No & Life Insurance Promotion = No*

THEN *Credit Card Insurance = No (4/4)*

IF *Watch Promotion = No*

THEN *Life Insurance Promotion = No & Credit Card Insurance = No (4/6)*

IF *Credit Card Insurance = No*

THEN *Watch Promotion = No & Life Insurance Promotion = No (4/8)*

Exercises at the end of the chapter ask you to write additional association rules for this dataset.

7.3.3 General Considerations

Association rules are particularly popular because of their ability to find relationships in large databases without having the restriction of choosing a single dependent variable. However, caution must be exercised in the interpretation of association rules since many discovered relationships turn out to be trivial.

As an example, let's suppose we present a total of 10,000 transactions for a market basket analysis. Also, suppose 70% of all transactions involve the purchase of milk and 50% of all transactions have a bread purchase. From this information, we are likely to see an association rule of the form:

> *If customers purchase milk they also purchase bread.*

The support for this rule may be well above 40% but note that most customers purchase both products. The rule does not give us additional marketing information telling us that it would be to our advantage to promote the purchase of bread with milk, since most customers buy these products together anyway. However, there are two types of relationships found within association rules that are of interest:

- We are interested in association rules that show a lift in product sales for a particular product where the lift in sales is the result of its association with one or more other products. In this case, we can use this information to help promote the product with increased sales as a result of the association.

- We are also interested in association rules that show a lower than expected confidence for a particular association. In this case, a possible conclusion is the products listed in the association rule compete for the same market.

As a final point, huge volumes of data are often stored for market basket analysis. Therefore, it is important to minimize the work required by an association rule generator. A good scenario is to specify an initially high value for the item set coverage criterion. If more rules are desired, the coverage criterion can be lowered and the entire process repeated. It's time to experiment!

7.3.4 Rweka's Apriori Function

R offers several association rule packages from which to choose. For our experiments here, we use *RWeka's Apriori* association rule function. We will look at the *apriori(arules package)* function in Section 7.4.

Most association rule generators including Apriori are limited to datasets containing nominal (categorical) data. If a dataset contains numeric attributes, we have one of two choices. One option is to delete all numeric attributes prior to generating association rules.

Our second option is to convert numeric attributes to categorical equivalents. For our first example, we examine the contact-lenses dataset included in the RWeka package. The dataset contains 24 instances, with four categorical input attributes, and an output attribute stating whether the individual is able to wear contact lenses. The dataset is stored in the standard Weka file *arff* format. The read statement in Script 7.4 reads and converts the file to a format suitable for the RStudio environment.

Script 7.4 shows the statements and relevant output seen when *Apriori* is applied to the contact lenses data. The *WOW* function lists the options available with *Apriori*. We can choose a lower bound for confidence, set lower and upper bounds on support, use a metric other than confidence, and control the maximum number of rules to be displayed. Script 7.4 limits the options list to those actually used here.

Script 7.4 Apriori Rules: Contact-Lenses Dataset

```
> library(RWeka)

> WOW(Apriori)   #Options
-N <required number of rules output>
        The required number of rules. (default = 10)
      Number of arguments: 1.
-M <lower bound for minimum support>
        The lower bound for the minimum support. (default = 0.1)
      Number of arguments: 1.

>contact.data <- read.arff(system.file("arff",
+ "contact-lenses.arff,package="RWeka"))


> # Set number of rules at 6
> Apriori(contact.data, control= Weka_control(N=6))

Apriori
=======

Minimum support: 0.2 (5 instances)
Minimum metric <confidence>: 0.9
Number of cycles performed: 16

Best rules found:

1. tear.prod.rate=reduced 12 ==> contact.lenses=none 12 <conf:(1)

2. spectacle.prescrip=hypermetrope tear.prod.rate=reduced 6 ==>
contact.lenses=none 6   <conf:(1)
```

```
3. spectacle.prescrip=myope tear.prod.rate=reduced 6 ==> contact.
lenses=none 6  <conf: (1)

4. astigmatism=no tear.prod.rate=reduced 6 ==> contact.lenses=none
6 <conf: (1)

5. astigmatism=yes tear.prod.rate=reduced 6 ==> contact.
lenses=none 6 <conf: (1)

6. contact.lenses=soft 5 ==> astigmatism=no 5 <conf: (1)

# Set the minimum support at 50%

> Apriori(contact.data, control= Weka_control(M=.5))

Apriori
======


Minimum support: 0.5 (12 instances)
Minimum metric <confidence>: 0.9
Number of cycles performed: 10

Best rules found:

  1. tear.prod.rate=reduced 12 ==> contact.lenses=none 12 <conf: (1)
```

The first call to *Apriori* generates six association rules. The first five rules display *contact-lenses* as the output attribute. *Astigmatism* is the output attribute in rule 6. Let's look at the second rule given as

If spectacle-prescrip = myope tear-prod-rate=reduced 6 → contact-lenses=none 6 conf:(1)

The 6 in the rule precondition (left of → symbol) tells us that six data instances show both myope and reduced as values for the given attributes. The 6 in the rule consequent (right of → symbol) tells us that these same six instances have the value *none* for the attribute contact-lenses. This results in a confidence value of 1 for the rule. All listed rules have a confidence score of 1 meaning that each rule is 100% accurate.

The *Apriori* function does not display rule support. However, the *WOW* option list tells us LowerBoundMinSupport is currently set at 0.1. If we reset this value at 50% (M= 0.5) and run the application a second time, we see but a single rule. Specifically,

Tear-prod-rate= reduced → contact-lenses = none

Let's turn our attention to a more interesting dataset designed for market basket analysis. The dataset—*supermarket.arff*—contains actual shopping data collected from a supermarket in New Zealand. This dataset is not part of RWeka but is included in your supplementary materials. Script 7.5 shows how the *read.arff* function reads the dataset from the current working directory.

Script 7.5 Apriori Rules: Supermarket Data

```
> library(RWeka)
> super.data <-read.arff("supermarket.arff")
> super.ap <- Apriori(super.data, control= Weka_control(C = .9,
N=5))
> super.ap


Apriori
======

Minimum support: 0.15 (694 instances)
Minimum metric <confidence>: 0.9
Number of cycles performed: 17


1. biscuits=t frozen foods=t fruit=t total=high 788 ==>
bread and cake=t 723      <conf:(0.92)
2. baking needs=t biscuits=t fruit=t total=high 760 ==>
bread and cake=t 696      <conf:(0.92)
3. baking needs=t frozen foods=t fruit=t total=high 770 ==>
bread and cake=t 705 <conf:(0.92)
4. biscuits=t fruit=t vegetables=t total=high 815 ==>
bread and cake=t 746      <conf:(0.92)
5. party snack foods=t fruit=t total=high 854 ==>
bread and cake=t
779 < conf:(0.91)
```

The entire dataset contains 4627 instances and 217 attributes. Each instance gives the purchases of a shopper in a single trip to the supermarket. Figure 7.1 displays an RStudio screenshot showing a small segment of the data. *View(super.data)* will give you a better overall picture of the attribute names and values.

The original *.arff* file holds a wealth of question marks where each ? represents an item not purchased. Fortunately, the *read.arff* function converts all ?'s to NA's!

Some attributes are very general in that they simply indicate a purchase within a given department. Other attributes are more specific as their value is an individual item type such as canned fruit. The last attribute is *total* with a value of *high* or *low*. The value is high if the total bill is more than $100. However, this attribute is not necessarily of major importance as we are looking for association rules featuring combinations of attribute values in order to determine those items likely to be purchased together.

Each purchase is represented by a *t*. The first listed customer purchased a *baby needs* item, a *bread and cake* item, an item from *baking needs* and several other items seen by

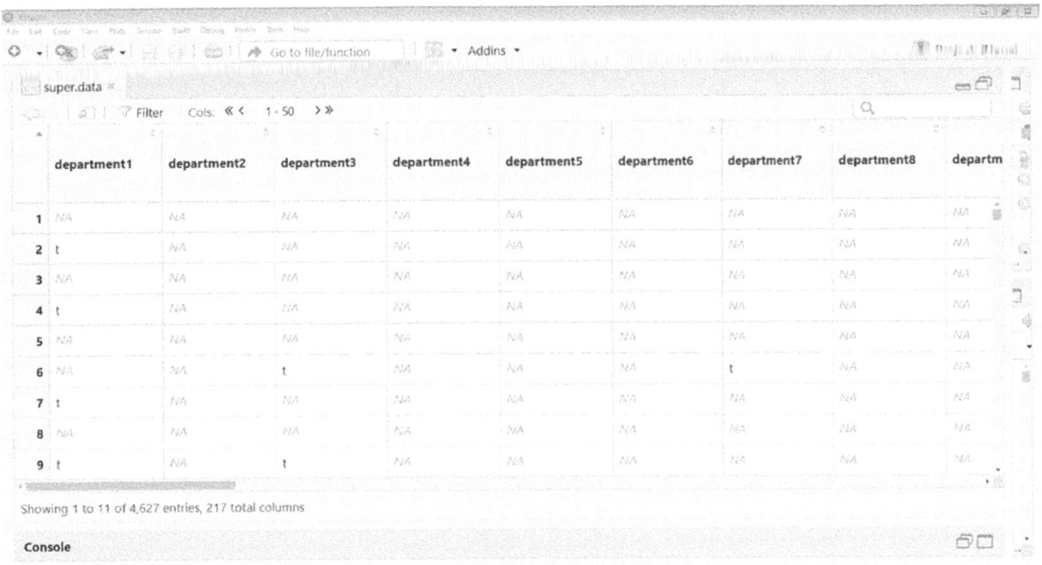

FIGURE 7.1 The supermarket dataset.

moving the scroll bar at the bottom of the edit window or clicking the *cols* arrows at the top of the screen. Customer 2 purchased an item from *department 1* as well as the additional items seen by moving the scroll bar to the right. It is obvious that most of the data is quite sparse in that most attribute values are empty.

Using the Script 7.5 settings for the *Apriori* function produces five association rules. It is clear that the consequent for each rule is *bread and cake*. Further investigation shows that all rules generated with the default settings have the same consequent! To find other rules of possible interest requires manipulation of one or several parameter settings. Simply modifying the minimum confidence requirement to 70% ($C = 0.70$) brings several new attributes into the mix. End-of-chapter Exercise 10 challenges you to further experiment with these data in an attempt to find meaningful associations. But now, get ready to shake, *rattle*, and roll!

7.4 SHAKE, RATTLE, AND ROLL

In this section, we introduce Rattle (R Analytic Tool to Learn Easily), a GUI that allows us to visualize data, preprocess data, build models, and evaluate results without writing lengthy scripts! First, we must head to the CRAN library to install and load the *rattle package*.

Once the *rattle package* is installed and loaded, you launch the GUI by typing *rattle( )* in the console window. Figure 7.2 displays the interface. With Rattle, we can load, explore, transform, cluster, and model data. Rattle also contains a wealth of tools for model evaluation.

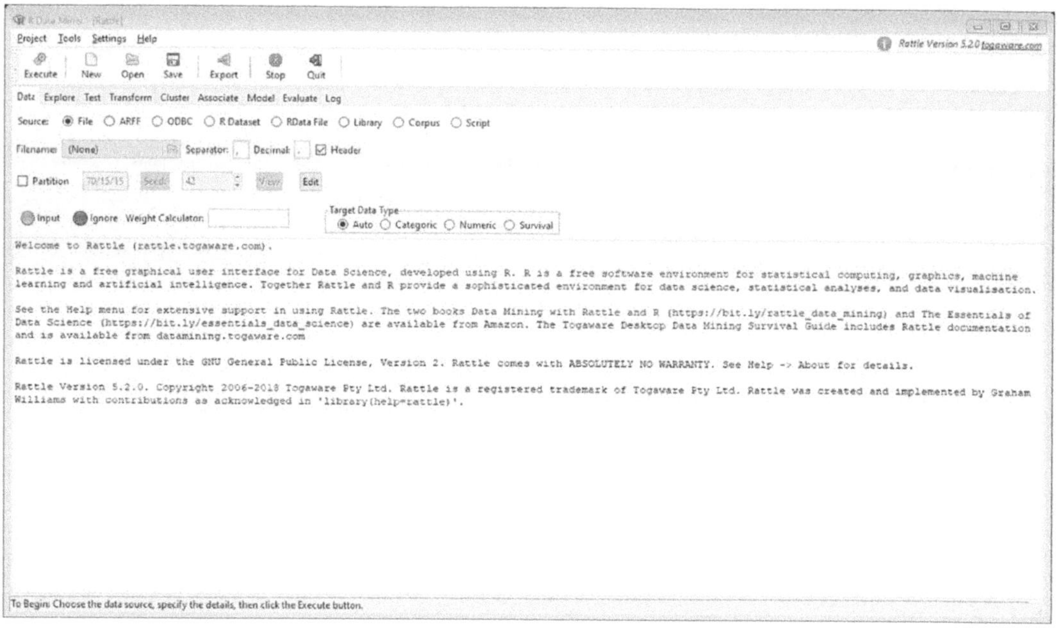

FIGURE 7.2   The Rattle interface.

The most efficient way for you to learn how to use Rattle is to work through a few examples. As we work through these examples, you will soon learn that Rattle requires several packages that are not installed on your machine. Each time Rattle asks you if a missing package can be installed, simply give an affirmative answer and let Rattle do the work. Let's start simple by applying *rpart* to the *contact-lenses* dataset.

Example 1: A Decision Tree Application

We start by making the *contact-lenses* dataset available to Rattle by typing the following statement into the RStudio console:

>*contact.data <- read.arff(system.file("arff", "contact-lenses.arff", package = "RWeka"))*

Next, return to the Rattle interface and highlight the *R Dataset* button. Scroll the *Data Name* window to find *contact.data*. Click the *execute* button in the upper left of your screen. This loads the data into the Rattle environment. Figure 7.3 shows the result.

We want to build but not evaluate a decision tree. If necessary, deselect *partition* and click *execute* to finalize the change. Make sure *contact.lenses* is specified as the target variable. Next, click on *model* located right below the red *stop* button. Select type *Tree*, algorithm *traditional*, and change *Min Split* to 5. Click *execute* to display the statements defining the tree as in Figure 7.4. Click *rules* on the right side of your screen to see the rules generated for the tree. Figure 7.5 offers a graphical depiction of the decision tree obtained by clicking on *draw*.

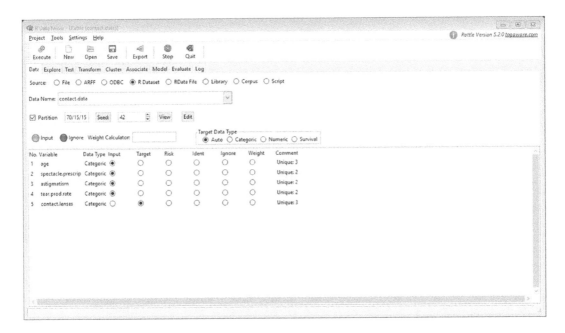

FIGURE 7.3   Loading the contact-lenses dataset into Rattle.

FIGURE 7.4   The statements defining a decision tree for the contact-lenses dataset.

FIGURE 7.5   An rpart decision tree for the contact-lenses dataset.

Example 2: A Random Forest Application

Our second example is a random forest application using the customer churn data-set. To begin, *click data → R Dataset*. Scroll to find *churnTrain* and click *execute*. Deselect *partition* and click *execute* as we have an available test dataset. Your screen will appear as in Figure 7.6. The first listed variable given as *state* is categorical with 51 unique values. Select *ignore* for this variable as random forest can only handle categorical attributes with 32 or fewer unique factor values. Click *execute* to effect the change.

Next, go to model and select Forest. Using all default values, click execute. Your screen will appear as in Figure 7.7. Each tree within the forest is built using a bootstrap technique, thereby leaving 1/3 of the training data available for testing.

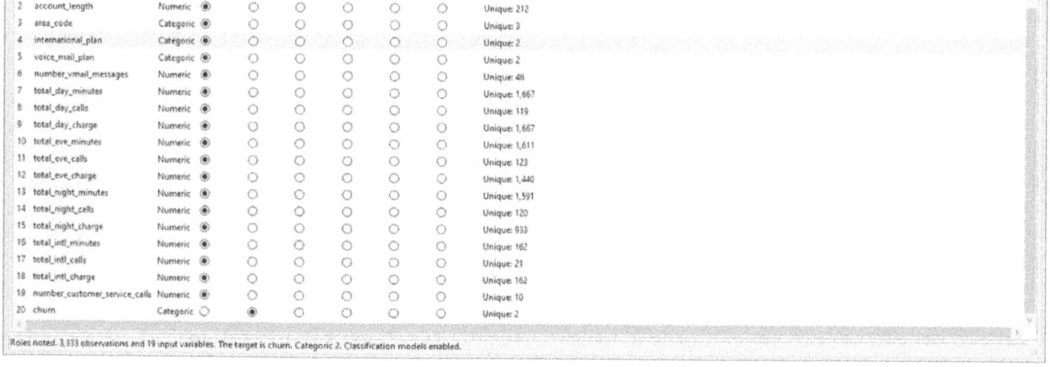

FIGURE 7.6   Churn training data loaded into Rattle.

The out-of-bag (OOB) error estimate shown in Figure 7.7 as 4.44% is based on how each tree performs when presented with their OOB data. The confusion matrix represents the performance of the forest when given the entire set of training data.

Figure 7.7 also displays options for *Rules*, *Errors*, and *OOB ROC*. Click on *Errors* to have Figure 7.8 appear in the *plots* area of your RStudio screen. The graph tells us that the OOB error rate will likely remain the same even if we remove over half of the trees. The *OOB ROC* curve (not shown) displays an impressive *AUC* of 0.913.

To see how our model does with the test data, click on *Evaluate* (just right of *model*). Highlight the *R Dataset* radio button, scroll to locate *churnTest* and click *execute*. Figure 7.9 displays the confusion matrix resulting from applying the forest model

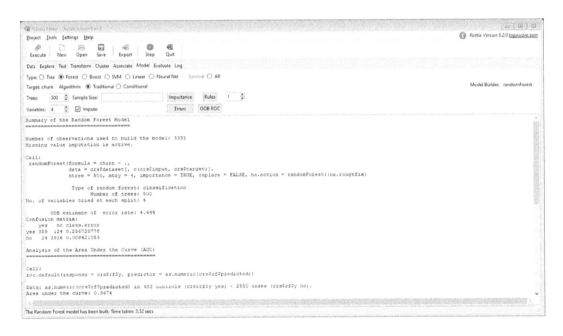

FIGURE 7.7 A Random Forest for the churnTrain dataset.

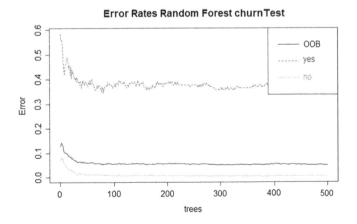

FIGURE 7.8 Error rates for customer churn training data.

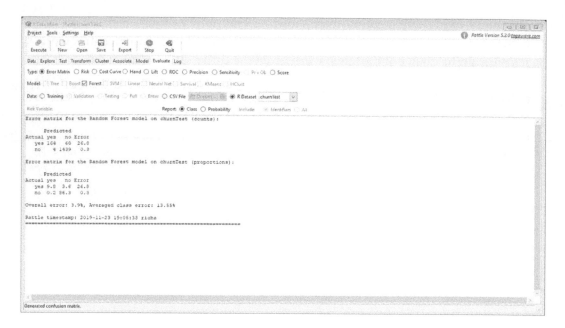

FIGURE 7.9   Test set error rates for customer churn data.

to the test data. The false positive rate of 0.278% is impressive! This tells us we won't be spending a lot on incentives for non-churners. On the negative side, we still miss almost 27% of the folks that churn. If the identification of a larger percent of churners is a main goal, experimenting with the number of trees in the forest and/or the number of attributes is a good starting point.

Example 3: Mining Association Rules

Rattle interfaces with the *apriori* association rule function that comes with the *arules package*. Let's see what kind of associations it can find in the *UCI Adult dataset*. The dataset contains census information for 48,842 individuals and is formatted for use with arules. It's worth your time to read more about the nature of the data prior to performing the experiment below. To learn more, click on the *Adult* link within the *arules* package.

Here's how to access the data through the *rattle* interface:

- Click *data* then *library*.
- Scroll to find *AdultUCI:arules:Adult Data Set*. Click *execute* to load the data.
- Figure 7.10 lists the attributes within the data. Make *income* an input attribute and deselect *partition*, click *execute*.
- Click *associate*, set the value of support at 0.4 and confidence 0.80, and click *execute*.
- Click on *show rules*. Figure 7.11 displays the association rules.

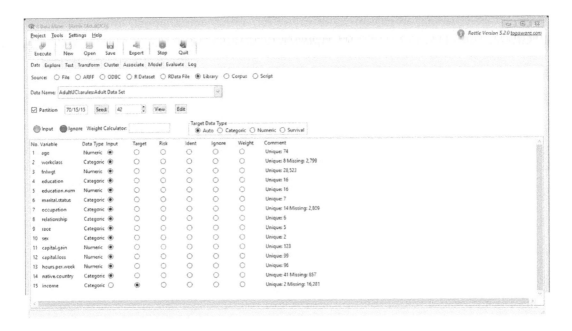

FIGURE 7.10   The AdultUCI dataset.

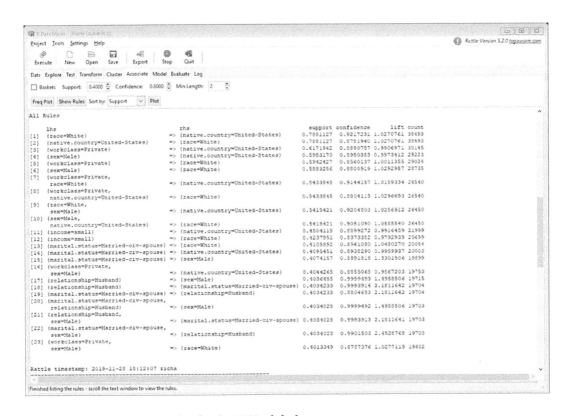

FIGURE 7.11   Association rules for the UCI adult dataset.

Each listed association rule includes values for confidence, support, lift, and count. Given a rule of the form $x \rightarrow y$, the equation for computing *lift* is

$$\frac{support(x,y)}{support(x)^* support(y)}$$

where *support(x,y)* is the value of rule support.

Values for lift above 1.0 increase the likelihood of a purchase of *y* given a purchase of *x*, values below 1.0 indicate the purchase of *x* decreases the likelihood of purchasing *y*. Although the rules displayed in Figure 7.11 are of little interest, the dataset is large and worthy of further experimentation.

Lastly, you can generate rules using the *arules* function for datasets directly in your console. Here's one possibility for the current data.

```
library(arules)
data("Adult")
my.Rules <- apriori(Adult,
+ parameter = list(supp = 0.4, conf = 0.8, target = "rules"))
summary(my.Rules)
inspect(my.Rules)
```

7.5 CHAPTER SUMMARY

Decision trees map nicely to a set of production rules by writing one rule for each path of the tree. In some cases, the mapped rules can be simplified without a significant loss in test set accuracy levels.

Covering rule generators attempt to create a set of rules that maximize the total number of covered within-class instances while at the same time minimizing the number of covered nonclass instances. Association rules are unlike traditional production rules in that an attribute that is a precondition in one rule may appear as a consequent in another rule. Also, association rule generators allow the consequent of a rule to contain one or several attribute values. As association rules are more complex, special techniques have been developed to generate association rules efficiently. Rule confidence and support help determine which discovered associations are likely to be interesting from a marketing perspective. However, caution must be exercised in the interpretation of association rules because many discovered relationships turn out to be trivial. The *rattle* package offers us an easy-to-use GUI for exploring, transforming, clustering, and modeling data.

7.6 KEY TERMS

- *Affinity analysis.* The process of determining which things are typically grouped together.

- *Confidence.* Given a rule of the form "If A then B," confidence is defined as the conditional probability that B is true when A is known to be true.

- *Item set.* A set of several combinations of attribute-value pairs that cover a prespecified minimum number of data instances.

- *Support.* The minimum percentage of instances in the database that contain all items listed in a given association rule.

EXERCISES

Review Questions

1. How do association rules differ from traditional production rules? How are they the same?

2. How do covering rules differ from traditional production rules?

Experimenting with R

1. Use C5.0 (default settings) to generate rules for the ccpromo dataset with LifeInsPromo as the output attribute. Repeat but use *JRip* to generate the rules. Compare the two sets of rules. Summarize your results.

2. Experiment with various values for *minCases* in Script 7.1. Attempt to create a model with fewer rules and at the same time maintain overall classification accuracy. Make a table to show the results of your experiments. How does the value for *minCases* affect the number of valid emails classified as spam?

3. Use Script 7.2 to experiment with the spam email cutoff value. Make a table showing the overall classification accuracy, the number of valid emails classified as spam, and the number of spam emails classified as valid for various cutoff values. What value should be used if the goal is to have less than 2% of all valid emails going to the junk box?

4. Use the tables created in Exercises 2 and 3 to help you experiment with various combinations for *minCases* and cutoff values in order to obtain a best result.

5. Repeat the experiment in Script 7.1, but replace C5.0 with RWeka's PART rule generator. Be sure to include the RWeka library in your script. Use *WOW(PART)* to learn about the options available with PART. The initial statement for building your model will be of the form:

   Spam.PART <- PART(Spam~., data = spam.train, control = Weka_control(M=2))

   Experiment with higher values for *M* to generate fewer rules. Attempt to build a model that reduces misclassifications of valid emails without compromising overall accuracy.

6. Repeat the experiments in Scripts 7.1 and 7.2 using the *churn* data introduced in Chapter 6. The goal is to build a model that identifies the largest percent of test set churn candidates but at the same time maintains a reasonable overall test set accuracy. Start by defining a specific goal.

7. Apply Rattle's association rule generator to the ccpromo dataset. Specify rule confidence as 0.90 and minimum support at 0.40. State what you consider to be the two most interesting rules and why they are of interest.

8. Apply Rattle's association rule generator to the contact-lenses dataset. Specify rule confidence as 0.90 and minimum support at 0.10. State what you consider to be the two most interesting rules and why they are of interest.

9. Write a script using *apriori(arules)* to generate rules for the ccpromo dataset. As *age* is numeric, you must either discretize or delete *age* prior to rule generation. Experiment with various settings for support and confidence. Report on two or three "interesting" rules found within the data.

10. Experiment with RWeka's implementation of *Apriori* and the supermarket data. Modify confidence and support values in an attempt to find interesting rules. Comment on three of the rules you consider to be interesting. Specify the confidence and support values and why the rules are of interest.

11. Use *JRip* to create a covering rule set for the *Spam* dataset. Experiment with *JRip's* parameters to obtain a best result with a minimum number of rules. Do a comparative analaysis of the rules created by C50, PART, and JRip.

12. Use *PART* to generate rules for the *churn* dataset. Compare the test set accuracy of your model with the results seen when *JRip* was applied to these same data.

Computational Questions

1. Verify the accuracy and lift values for the following rule generated by C5.0 using all instances of the ccpromo dataset.

```
Rule 1:  (6/1, lift 1.9)
         CCardIns = No
         Gender = Male
         -> class No   [0.750]
```

2. Answer the following.

a. Write the production rules for the decision tree shown in Figure 6.3.

b. Repeat the previous question for the decision tree shown in Figure 6.5.

3. Use the data in Table 7.1 to give confidence and support values for the following association rule.

If Gender = Male & Magazine Promotion = Yes then Life Insurance Promotion = Yes

4. Use the information in Table 7.3 to list three two-item set rules. Use the data in Table 7.1 to compute confidence and support values for each of your rules.

5. List three rules for the following three-item set. Use the data in Table 7.1 to specify the confidence and support for each rule.

   Watch Promotion = No & Life Insurance Promotion = No & Credit Card Insurance = No

Installed Packages and Functions

Package Name	Function(s)
arules	apriori, inspect
base/ stats	c, cbind, data.frame, factor, ifelse, library, predict set.seed, sample, summary, table, View
C50	C5.0
randomForest	randomForest
rattle	rattle
rpart	rpart
RWeka	Apriori, JRip, PART, read.arff, WOW

CHAPTER **8**

Neural Networks

In This Chapter

- Feed-Forward Neural Networks

- Self-Organizing Networks

- Strengths and Weaknesses of Neural Networks

- Building Neural Networks with R

NEURAL NETWORKS CONTINUE TO grow in popularity within the business, scientific, and academic worlds. This is because neural networks have a proven track record in predicting both continuous and categorical outcomes. An excellent overview of neural network applications is given in Widrow et al. (1994).

Although several neural network architectures exist, we limit our discussion to two of the more popular structures. For supervised classification, we examine feed-forward neural networks trained with backpropagation. For unsupervised clustering, we discuss Kohonen self-organizing maps (SOMs).

Section 8.1 introduces you to some of the basic concepts and terminology for neural networks. Neural networks require numeric input values ranging between 0 and 1 inclusive. As this can be a problem for some applications, we discuss neural network input and output issues in detail. In Section 8.2, we offer a conceptual overview of how supervised and unsupervised neural networks are trained. Neural networks have been criticized for their inability to explain their output. Section 8.3 looks at some of the techniques that have been developed for neural network explanation. Section 8.4 offers a list of general strengths and weaknesses found with all neural networks. Section 8.5 presents detailed examples of how backpropagation and self-organizing neural networks are trained. If your interests do not lie in a precise understanding of how neural networks learn, you may want to skip Section 8.5.

In Section 8.6, you will use two packages available with R to create and test supervised neural net models. Section 8.7 is all about unsupervised neural net clustering with R. The focus of the last section is on stock market prices and time series data.

8.1 FEED-FORWARD NEURAL NETWORKS

Neural networks offer a mathematical model that attempts to mimic the human brain. Knowledge is often represented as a layered set of interconnected processors. These processor nodes are frequently referred to as *neurodes* so as to indicate a relationship with the neurons of the brain. Each node has a weighted connection to several other nodes in adjacent layers. Individual nodes take the input received from connected nodes and use the connection weights together with a simple evaluation function to compute output values.

Neural network learning can be supervised or unsupervised. Learning is accomplished by modifying network connection weights while a set of input instances is repeatedly passed through the network. Once trained, an unknown instance passing through the network is classified according to the value(s) seen at the output layer.

Figure 8.1 shows a fully connected *feed-forward* neural network structure together with a single input instance [1.0, 0.4, 0.7]. Arrows indicate the direction of flow for each new instance as it passes through the network. The network is *fully connected* because nodes at one layer are connected to all nodes in the next layer.

The number of input attributes found within individual instances determines the number of input layer nodes. The user specifies the number of hidden layers as well as the number of nodes within a specific hidden layer. Determining a best choice for these values is a matter of experimentation. In practice, the total number of hidden layers is usually restricted to two. Depending on the application, the output layer of the neural network may contain one or several nodes.

8.1.1 Neural Network Input Format

The input to individual neural network nodes must be numeric and fall in the closed interval range [0, 1]. Because of this, we need a way to numerically represent

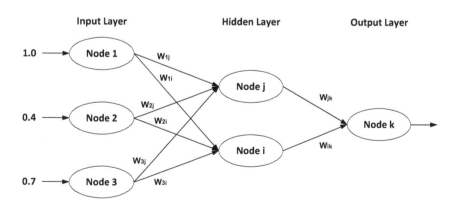

FIGURE 8.1 A fully connected feed-forward neural network.

categorical data. We also require a conversion method for numerical data falling outside the [0, 1] range.

There are several choices for categorical data conversion. A straightforward technique divides the interval range into equal-size units. To illustrate, consider the attribute *color* with possible values of red, green, blue, and yellow. Using this method, we might make the assignments: red = 0.00, green = 0.33, blue = 0.67, and yellow = 1.00. Although this technique can be used, it has an obvious pitfall. The modification incorporates a measure of distance not seen prior to the conversion. This is shown in our example in that the distance between red and green is less than the distance between red and yellow. Therefore, it appears as though the color red is more similar to green than it is to yellow.

A second technique for categorical to numerical conversion requires one input node for each categorical value. Once again, consider the attribute *color* with the four values assigned as in the previous example. By adding three additional input nodes for *color*, we can represent the four colors as follows: red = [1,0,0,0], green = [0,1,0,0], blue = [0,0,1,0], and yellow = [0,0,0,1]. Using this scheme, each categorical value becomes an attribute with possible values 0 or 1. It is obvious that this method eliminates the bias seen with the previous technique.

Now let's consider the conversion of numerical data to the required interval range. Suppose we have the values 100, 200, 300, and 400. An obvious conversion method is to divide all attribute values by the largest attribute value. For our example, dividing each number by 400 gives the converted values: 0.25, 0.5, 0.75, and 1.0. The problem with this method is that we cannot take advantage of the entire interval range unless we have at least some values close to zero. A slight modification of this technique offers the desired result, as shown in Equation 8.1.

$$newValue = \frac{originalValue - minimumValue}{maximumValue - minimumValue} \tag{8.1}$$

where
  newValue is the computed value falling in the [0,1] interval range
  originalValue is the value to be converted
  minimumValue is the smallest possible value for the attribute
  maximumValue is the largest possible attribute value

Applying the formula to the values above gives us 0.0, 0.33, 0.66, and 1.0.

A special case exists when maximum values cannot be determined. One possible solution is to use an arbitrarily large value as a divisor. Once again, dividing by an arbitrary number leaves us with the possibility of not covering the entire interval range. Finally, highly skewed data may cause less than optimal results unless variations of these techniques are applied. A common approach with skewed data is to take the base 2 or base 10 logarithm of each value before applying one of the previous transformations.

8.1.2 Neural Network Output Format

The output nodes of a neural network represent continuous values in the [0, 1] range. However, the output can be transformed to accommodate categorical class values. To illustrate, suppose we wish to train a neural network to recognize new credit card customers likely to take advantage of a special promotion. We design our network architecture with two output layer nodes, nodes 1 and 2. During training, we indicate a correct output for customers that have taken advantage of previous promotions as *1* for the first output node and *0* for the second output node. A correct output for customers that traditionally do not take advantage of promotions is designated by a node 1, node 2 combination of *0* and *1,* respectively. Once trained, the neural network will recognize a node 1, node 2 output combination of 0.9, 0.2 as a new customer likely to take advantage of a promotion.

This method has certain disadvantages in that node output combinations such as 0.2, 0.3 have no clear classification. Various approaches have been proposed for this situation. One method suggests the association of certainty factors with node output values. A popular method uses a special test dataset to help with difficult-to-interpret output values. This method also allows us to build a neural network with a single output layer node even when the output is categorical. An example illustrates the method.

Suppose we decide on a single output node for the credit card customer example just discussed. We designate *1* as an ideal output for customers likely to take advantage of a special promotion and a *0* for customers likely to pass on the offer. Once we have trained the network, we can be confident about classifying an output value of 0.8 as a customer likely to take advantage of the promotion. However, what do we do when the output value is 0.45? The special test dataset helps with our dilemma. Prior to applying the network to unknown instances, we present the test set to the trained network and record the output values for each test instance. We then apply the network to the unknown instances. When unknown instance x shows an uncertain output value v, we classify x with the category shown by the majority of test set instances clustering at or near v.

Finally, when we wish to use the computed output of a neural network for prediction, we have another problem. Let's assume a network has been trained to help us predict the future price of our favorite stock. As the output of the network gives a result between 0 and 1, we need a method to convert the output into an actual future stock price.

Suppose the actual output value is 0.35. To determine the future stock price, we need to undo the original [0, 1] interval conversion. The process is simple. We multiply the training data range of the stock price by 0.35 and add the lowest price of the stock to this result. If the training data price range is $10.00 to $100.00, the computation is

$$(90.00)(0.35) + \$10.00$$

This gives a predicted future stock price of $41.50. All commercial and some public domain neural network packages perform these numeric conversions to and from the [0, 1] interval range. This still leaves us with the responsibility of making sure all initial inputs are numeric.

8.1.3 The Sigmoid Evaluation Function

The purpose of each node within a feed forward neural network is to accept input values and pass an output value to the next higher network layer. The nodes of the input layer pass input attribute values to the hidden layer unchanged. Therefore, for the input instance shown in Figure 8.1, the output of node 1 is 1.0, the output of node 2 is 0.4, and the output of node 3 is 0.7.

A hidden or output layer node n takes input from the connected nodes of the previous layer, combines the previous layer node values into a single value, and uses the new value as input to an evaluation function. The output of the evaluation function is the output of the node, which must be a number in the closed interval [0, 1].

Let's look at an example. Table 8.1 shows sample weight values for the neural network of Figure 8.1. Consider node j. To compute the input to node j, we determine the sum total of the multiplication of each input weight by its corresponding input layer node value. That is,

$$\text{Input to node } j = (0.2)(1.0) + (0.3)(0.4) + (-0.1)(0.7) = 0.25$$

Therefore, 0.25 represents the input value for node j's evaluation function.

The first criterion of an evaluation function is that the function must output values in the [0, 1] interval range. A second criterion is that as x increases, $f(x)$ should output values close to 1. In this way, the function propagates activity within the network. The *sigmoid function* meets both criteria and is often used for node evaluation. The sigmoid function is computed as

$$f(x) = \frac{1}{1 + e^{-x}} \tag{8.2}$$

where

e is the base of natural logarithms approximated by 2.718282.

Figure 8.2 shows the graph of the sigmoid function. Notice values of x less than zero provide little output activation. For our example, $f(0.25)$ evaluates to 0.562, which represents the output of node j.

Finally, it is important to note that a *bias* or *threshold* node is oftentimes associated with each hidden and output layer node. Threshold nodes differ from regular nodes in that their link values change during the learning process in the same manner as all other network weights, but their input is fixed at a constant value (usually 1). Later in this chapter, the purpose and functioning of threshold nodes is made clear when we investigate one of R's neural network functions.

TABLE 8.1   Initial Weight Values for the Neural Network Shown in Figure 8.1

W_{ij}	W_{li}	W_{2j}	W_{2i}	W_{3j}	W_{3i}	W_{jk}	W_{ik}
0.20	0.10	0.30	−0.10	−0.10	0.20	0.10	0.50

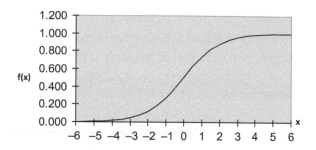

FIGURE 8.2   The sigmoid evaluation function.

8.2 NEURAL NETWORK TRAINING: A CONCEPTUAL VIEW

In this section, we discuss two methods for training feed-forward networks and one technique for unsupervised neural net clustering. Our discussion is limited in that we do not detail how the algorithms work. For most readers, the discussion here is enough to satisfy basic curiosities about neural network learning. However, for the more technically inclined individual, Section 8.5 offers specific details about how neural networks learn.

8.2.1 Supervised Learning with Feed-Forward Networks

Supervised learning involves both training and testing. During the training phase, training instances are repeatedly passed through the network while individual weight values are modified. The purpose of changing the connection weights is to minimize training set error between predicted and actual output values. Network training continues until a specific terminating condition is satisfied. The terminating condition can be convergence of the network to a minimum total error value, a specific time criterion, or a maximum number of iterations.

Backpropagation learning is most often used to train feed-forward networks. For each training instance, backpropagation works by first feeding the instance through the network and computing the network output value. Recall that one output value is computed for each output layer node. To illustrate backpropagation learning, we use Figure 8.1 together with the input instance shown in the figure.

We previously determined that the computed output for the instance in Figure 8.1 is 0.52. Now suppose that the target output associated with the instance is 0.65. Obviously the absolute error between the computed and target value is 0.13. However, a problem is seen when we attempt to determine why the error occurred. That is, we do not know which of the network connection weights is to be blamed for the error. It is possible that changing just one of the weights will provide us with a better result the next time the instance passes through the network. It is more likely that the problem lies with some combination of two or more weight values. Still another possibility is that the error is to some degree the fault of every network connection associated with the output node.

The backpropagation learning algorithm assumes this last possibility. For our example, the output error at node k is propagated back through the network, and all eight of the associated network weights change value. The amount of change seen with each connection weight is computed with a formula that makes use of the output error at node k, individual node output values, and the derivative of the sigmoid function. The formula has a way of smoothing the actual error value so as not to cause an overcorrection for any one training instance.

Given enough iterations, the backpropagation learning technique is guaranteed to converge. However, there is no guarantee that the convergence will be optimal. Therefore, several applications of the algorithm may be necessary to achieve an acceptable result.

8.2.2 Unsupervised Clustering with Self-Organizing Maps

Teuvo Kohonen (1982) first formalized neural network unsupervised clustering in the early 1980s when he introduced Kohonen feature maps. His original work focused on mapping images and sounds. However, the technique has been effectively used for unsupervised clustering. Kohonen networks are also known as self-organizing maps (SOMs).

Kohonen networks support two layers. The input layer contains one node for each input attribute. Nodes of the input layer have a weighted connection to all nodes in the output layer. The output layer can take any form but is commonly organized as a two-dimensional grid. Figure 8.3 shows a simple Kohonen network with two input layer and nine output layer nodes.

During network learning, input instances are presented to each output layer node. When an instance is presented to the network, the output node whose weight connections most closely match the input instance *wins* the instance. The node is rewarded by having its weights changed to more closely match the instance. At first, neighbors to the winning node are also rewarded by having their weight connections modified to more closely match the attribute values of the current instance. However, after the instances have passed through the network several times, the size of the neighborhood decreases until finally only the winning node gets rewarded.

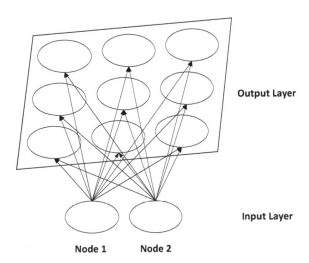

FIGURE 8.3 A 3 × 3 A Kohonen network with two input layer nodes.

Each time the instances pass through the network, the output layer nodes keep track of the number of instances they win. The output nodes winning the most instances during the last pass of the data through the network are saved. The number of output layer nodes saved corresponds to the number of clusters believed to be in the data. Finally, those training instances clustered with deleted nodes are once again presented to the network and classified with one of the saved nodes. The nodes, together with their associated training set instances, characterize the clusters in the dataset. Alternatively, test data may be applied, and the clusters formed by these data are then analyzed to help determine the meaning of what has been found.

8.3 NEURAL NETWORK EXPLANATION

A major disadvantage seen with the neural network architecture is a lack of understanding about what has been learned. Here we summarize four techniques for addressing this disadvantage. One possibility for neural network explanation is to transform a network architecture into a set of rules. Algorithms designed to extract rules from a neural network typically involve removing weighted links that minimally affect classification correctness. Unfortunately, rule extraction methods have met with limited success.

Sensitivity analysis is a second technique that has been successfully applied to gain insight into the effect individual attributes have on neural network output. There are several variations to the approach. The general process consists of the following steps:

1. Divide the data into a training set and a test dataset.

2. Train the network with the training data.

3. Use the test set data to create a new instance *I*. Each attribute value for *I* is the average of all attribute values within the test data.

4. For each attribute,

   a. Vary the attribute value within instance *I* and present the modification of *I* to the network for classification.

   b. Determine the effect the variations have on the output of the neural network.

   c. The relative importance of each attribute is measured by the effect of attribute variations on network output.

A sensitivity analysis allows us to determine a rank ordering for the relative importance of individual attributes. However, the approach does not offer an explicit set of rules to help us understand more about what has been learned.

A third technique that has merit as a generalized explanation tool for unsupervised clustering is the *average member technique*. With this method, the average or most typical member of each class is computed by finding the average value for each class attribute in each class (vs. the average value of all instances as done for sensitivity analysis).

A fourth and more informative alternative is to apply supervised learning to interpret the results of an unsupervised clustering. Here is the procedure as it applies to unsupervised neural network learning:

1. Perform any data transformations necessary to prepare the data for the unsupervised neural network clustering.

2. Present the data to the unsupervised network model.

3. Call each cluster created by the neural network a class and assign each cluster an arbitrary class name.

4. Use the newly formed classes as training instances for a decision tree algorithm.

5. Examine the decision tree to determine the nature of the concept classes formed by the clustering algorithm.

This method holds an advantage over the average member technique in that we are offered a generalization about differences as well as similarities of the formed clusters. As an alternative, a rule generator can be employed to describe each cluster in detail.

8.4 GENERAL CONSIDERATIONS

The process of building a neural network is both an art and a science. A reasonable approach is to conduct several experiments while varying attribute selections and learning parameters. The following is a partial list of choices that affect the performance of a neural network model:

- What input attributes will be used to build the network?

- How will the network output be represented?

- How many hidden layers should the network contain?

- How many nodes should there be in each hidden layer?

- What condition will terminate network training?

There are no right answers to these questions. However, we can use the experimental process to help us achieve desired results. Later in this chapter, you will learn how to better answer these questions by experimenting with R's neural net software tools. Here we provide a list of strengths and weaknesses for the neural network approach to knowledge discovery.

8.4.1 Strengths

- Neural networks work well with datasets containing large amounts of noisy input data. Neural network evaluation functions such as the sigmoid function naturally smoothen input data variations caused by outliers and random error.

- Neural networks can process and predict numeric as well as categorical outcome. However, categorical data conversions can be tricky.

- Neural networks have performed consistently well in several domains.

- Neural networks can be used for both supervised classification and unsupervised clustering.

8.4.2 Weaknesses

- Probably the biggest criticism of neural networks is that they lack the ability to explain their behavior.

- Neural network learning algorithms are not guaranteed to converge to an optimal solution. With most types of neural networks, this problem can be dealt with by manipulating various learning parameters.

- Neural networks can be easily trained to work well on the training data but poorly on test data. This often happens when the training data is passed through the network so many times that the network is not able to generalize when presented with previously unseen instances. This problem can be monitored by consistently measuring test set performance.

8.5 NEURAL NETWORK TRAINING: A DETAILED VIEW

Here we provide detailed examples of how two popular neural network architectures modify their weighted connections during training. In the first section, we provide a partial example of backpropagation learning and state a general form of the backpropagation learning algorithm. In the second section, we show you how Kohonen SOMs are used for unsupervised clustering.

8.5.1 The Backpropagation Algorithm: An Example

Backpropagation is the training method most often used with feed-forward networks. Backpropagation works by making modifications in weight values starting at the output layer and then moving backward through the hidden layers. The process is best understood with an example. We will follow one pass of the backpropagation algorithm using the neural network of Figure 8.1, the input instance shown in the figure, and the initial weight values from Table 8.1.

Let's assume the target output for the specified input instance is 0.68. The first step is to feed the instance through the network and determine the computed output for node k. We apply the sigmoid function to compute all output values, as shown in the following calculations.

Input to node $j = (0.2)(1.0) + (0.3)(0.4) + (-0.1)(0.7) = 0.250$

Output from node $j = 0.562$

Input to node $i = (0.1)(1.0)+(-0.1)(0.4)+(0.2)(0.7) = 0.200$

Output from node $i = 0.550$

Input to node $k = 0.1 * 0.562+0.5 * 0.550 = 0.331$

Output from node $k = 0.582$

Next we compute the observed error at the output layer of the network. The output layer error is computed as

$$Error(k) = (T - O_k)[f'(x_k)] \tag{8.3}$$

where
  $T =$ The target output
  $O_k =$ The computed output at node k
  $(T - O_k) =$ The actual output error
  $f'(x_k) =$ The first-order derivative of the sigmoid function
  $x_k =$ The input to the sigmoid function at node k.

Equation 8.3 shows that the actual output error is multiplied by the first-order derivative of the sigmoid function. The multiplication scales the output error, forcing stronger corrections at the point of rapid rise in the sigmoid curve. The derivative of the sigmoid function at x_k conveniently computes to $O_k(1 - O_k)$. Therefore,

$$Error(k) = (T - O_k)O_k(1 - O_k) \tag{8.4}$$

For our example, $Error(k)$ is computed as

$$Error(k) = (0.65 - 0.582)(0.582)(1 - 0.582) = 0.017$$

Computing the output errors for hidden layer nodes is a bit more intuitive. The general formula for the error at node j is

$$Error(j) = \left(\sum_k Error(k)W_{jk} \right) f'(x_j) \tag{8.5}$$

where
  $Error(k) =$ The computed output error at node k
  $W_{jk} =$ The weight associated with the link between node j and output node k
  $f'(x_j) =$ The first-order derivative of the sigmoid function
  $x_j =$ The input to the sigmoid function at node j. As in Equation 8.3, $f'(x_j)$ evaluates to

$$O_j(1-O_j)$$

Notice that the computed error is summed across all output nodes. For our example, we have a single output node. Therefore,

$$Error(j) = (0.017)(0.1)(0.562)(1-0.562) = 0.00042$$

We leave the computation of *Error(i)* as an exercise.

The final step in the backpropagation process is to update the weights associated with the individual node connections. Weight adjustments are made using the *delta rule* developed by Widrow and Hoff (Widrow and Lehr 1995). The objective of the delta rule is to minimize the sum of the squared errors, where error is defined as the distance between computed and actual output. We will give the weight adjustment formulas and illustrate the process with our example. The formulas are as follows:

$$w_{jk}(new) = w_{jk}(current) + \Delta w_{jk} \qquad (8.6)$$

where Δw_{jk} is the value added to the current weight value.

Finally, Δw_{jk} is computed as

$$\Delta w_{jk} = (r)[Error(k)](O_j) \qquad (8.7)$$

where

$r = $ The learning rate parameter with $1 > r > 0$

$Error(k) = $ The computed error at node k

$O_j = $ The output of node j.

Here are the parameter adjustments for our example with $r = 0.5$.

- $\Delta w_{jk} = (0.5)(0.017)(0.562) = 0.0048$
  The updated value for $w_{jk} = 0.1 + 0.0048 = 0.1048$

- $\Delta w_{1j} = (0.5)(0.00042)(1.0) = 0.0002$
  The updated value for $w_{1j} = 0.2 + 0.0002 = 0.2002$

- $\Delta w_{2j} = (0.5)(0.00042)(0.4) = 0.000084$
  The updated value for $w_{1j} = 0.3 + 0.000084 = 0.300084$

- $\Delta w_{3j} = (0.5)(0.00042)(0.7) = 0.000147$
  The updated value for $w_{1j} = -0.1 + 0.000147 = -0.099853$

We leave adjustments for the links associated with node *i* as an exercise. Now that we have seen how backpropagation works, we state the general backpropagation learning algorithm.

1. Initialize the network.
    a. Create the network topology by choosing the number of nodes for the input, hidden, and output layers.
    b. Initialize weights for all node connections to arbitrary values between –1.0 and 1.0.
    c. Choose a value between 0 and 1.0 for the learning parameter.
    d. Choose a terminating condition.
2. For all training set instances,
    a. Feed the training instance through the network.
    b. Determine the output error.
    c. Update the network weights using the previously described method.
3. If the terminating condition has not been met, repeat step 2.
4. Test the accuracy of the network on a test dataset. If the accuracy is less than optimal, change one or more parameters of the network topology and start over.

The terminating condition can be given as a total number of passes (also called *epochs*) of the training data through the network. Alternatively, network termination can be determined by the degree to which learning has taken place within the network. A generalized form of the *root mean squared error* is often used as a standard measure of network learning. The general formula to calculate *rms* is given as the square root of the following value.

$$\frac{\sum_n \sum_i (T_{in} - O_{in})^2}{ni} \tag{8.8}$$

where
$n =$ The total number of training set instances
$i =$ The total number of output nodes
$T_{in} =$ The target output for the *n*th instance and the *i*th output node
$O_{in} =$ The computed output for the *n*th instance and the *i*th output node.

As you can see, the *rms* is simply the square root of the average of all instance output error values. A common criterion is to terminate the backpropagation learning when the *rms* is less than 0.10. Variations of the just-stated approach exist. One common variation is to keep track of training data errors but wait to update network weight connections only after all training instances have passed through the network. Regardless of the methodology, several iterations of the process are often necessary for an acceptable result.

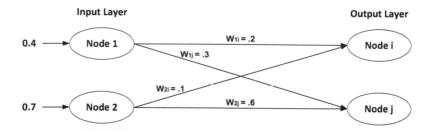

FIGURE 8.4   Connections for two output layer nodes.

8.5.2 Kohonen Self-Organizing Maps: An Example

To see how unsupervised clustering is accomplished, we consider the input layer nodes and two of the output layer nodes for the Kohonen feature map shown in Figure 8.3. The situation is displayed in Figure 8.4.

Recall that when an instance is presented to the network, a score for classifying the instance with each of the output layer nodes is computed. The score for classifying a new instance with output node j is given by

$$\sqrt{\sum_i (n_i - w_{ij})^2} \tag{8.9}$$

where n_i is the attribute value for the current instance at input node i, and w_{ij} is the weight associated with the ith input node and output node j. That is, the output node whose weight vectors most closely match the attribute values of the input instance is the winning node.

Let's use Equation 8.9 to compute the score for the two output nodes shown in Figure 8.4 using the input instance [0.4, 0.7]. The score for inclusion with output node i is

$$\sqrt{(0.4 - 0.2)^2 + (0.7 - 0.1)^2} = 0.632$$

Likewise, the score for inclusion with output node j is computed as

$$\sqrt{(0.4 - 0.3)^2 + (0.7 - 0.6)^2} = 0.141$$

As you can see, node j is the winner as its weight vector values are more similar to the input values of the presented instance. As a result, the weight vectors associated with the output node are adjusted so as to reward the node for winning the instance. The following formula is used to adjust the values of the weight vectors:

$$w_{ij}(new) = w_{ij}(current) + \Delta w_{ij}$$

where

$$\Delta w_{ij} = r\left(n_i - w_{ij}\right)$$

$$0 < r \leq 1$$

The winning output node has its weight vectors adjusted in the direction of the values contained within the new instance. For our example, with the learning rate parameter $r = 0.5$, the weight vectors of node j are adjusted as follows:

- $\Delta w_{1j} = (0.5)(0.4 - 0.3) = 0.05$
- $\Delta w_{2j} = (0.5)(0.7 - 0.6) = 0.05$
- $w_{1j}(new) = 0.3 + 0.05 = 0.35$
- $w_{2j}(new) = 0.6 + 0.05 = 0.65$.

Output layer nodes within a specified neighborhood of the winning node also have their weights adjusted using the same formula. A square grid typically defines the neighborhood. The center of the grid contains the winning node. The size of the neighborhood as well as the learning rate r is specified when training begins. Both parameters are decreased linearly over the span of several iterations. Learning terminates after a preset number of iterations or after instance classifications do not vary from one iteration to the next.

To complete the clustering, the output nodes have their connection weights fixed and all but the n most populated output nodes are deleted. After this, the original training data or a previously unseen test dataset is fed through the output layer one last time. If the original training data is used, those instances previously associated with a deleted node move to one of the appropriate remaining clusters. Finally, the clusters formed by the training or test data are analyzed possibly using one or both of the unsupervised evaluation methods described above in order to determine what has been discovered.

8.6 BUILDING NEURAL NETWORKS WITH R

With the basics in the rearview mirror, it's time to try our hand at building a few neural net models! R offers several packages with functions for training neural networks. Here we investigate two functions within the *RWeka package* and the *neuralnet* function found in the *neuralnet package*. You can easily install the neuralnet package as it is housed in the CRAN repository.

The RWeka package installation requires all little more work. However, as you will see, the neural network functions within RWeka are extremely useful in that they automatically take care of most preprocessing tasks and can be used with categorical input data. This extra effort is more than worth our time!

Script 8.1 provides the necessary code. We will be using the *MultiLayerPerceptron* function for backpropagation learning and the Kohonen *SelfOrganizingMap* function for unsupervised clustering. If you'd rather not take a chance at typing everything correctly, just load and execute the supplemental script code—*Script 8.1 Installing RWeka Network Packages.R*—using your source editor. Also, after exiting and reentering RStudio, you may have to reload the Kohonen unsupervised clustering algorithm (KSOM) and the MultiLayerPerceptron algorithm (MLP). A good way to determine if the functions are ready for use is to type MLP and KSOM in the console window. Doing so should output a summary statement about the specified function. If, instead you receive an error message try reloading one or both packages. This is accomplished by typing

>library(RWeka)

>WPM("load-packages", MLP,KSOM)

If this results in an error message, reinstall the packages by re-executing Script 8.1.

Script 8.1 Installing the RWeka Neural Network Packages

```
>library(RWeka)

WPM("refresh-cache")
WPM("install-package", "SelfOrganizingMap")
WPM("install-package", "multiLayerPerceptrons")
WPM("list-packages","installed")
KSOM<-make_Weka_clusterer("weka/clusterers/SelfOrganizingMap")
MLP<-make_Weka_classifier('weka/classifiers/functions/
MultilayerPerceptron')
KSOM
MLP
```

Once all packages are installed and loaded, you are ready to go! Let's start with MLP and the familiar XOR function. MLP offers a good starting point as it automatically converts any categorical input data to numeric equivalents, normalizes input data, and readily handles missing data.

8.6.1 The Exclusive-OR Function

Most of us are familiar with the basic logical operators. Common operators include *and, or, implication, negation,* and *exclusive or* (XOR). The definition of the XOR function is shown in Table 8.2. You can think of the XOR function as defining two classes. One class is denoted by the two instances with XOR function output equal to 1. The second class is given by the two instances with function output equal to 0.

Figure 8.5 offers a graphical interpretation of the output. The *x*-axis represents values for *X1*, and the *y*-axis denotes values for *X2*. The instances for the class with XOR equal 1 are denoted in Figure 8.5 by an *A*. Likewise, instances for the class with XOR equal to 0 are denoted with a *B*. The XOR function is of particular interest because, unlike the other

TABLE 8.2 Exclusive-OR Function

X1	X2	XOR
1	1	0
0	1	1
1	0	1
0	0	0

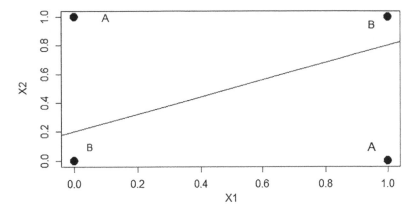

FIGURE 8.5   Graph of the XOR function.

logical operators, the points representing the XOR function are not *linearly separable*. This is seen in Figure 8.5, as we cannot draw a straight line to separate the instances in class A from those in class B.

The first neural networks, known as *perceptron networks*, consisted of an input layer and a single output layer. The XOR function caused trouble for these early networks because they converge only when presented with problems having linearly separable solutions. The development of the backpropagation network architecture, which is able to model nonlinear problems, contributed to a renewed interest in neural network technology.

Let's see how supervised models using feed-forward networks together with backpropagation learning are able to model the described XOR function. We'll follow a simple five-step approach:

1. Identify the goal.

2. Prepare the data.

3. Define the network architecture.

4. Watch the network train.

5. Read and interpret summary results.

Script 8.2 displays the training data for the XOR function where XOR is the output attribute.

Script 8.2 Training Data for the Exclusive-OR Function

```
> library(RWeka)
> X1<- c(0,1,0,1)
> X2<- c(0,0,1,1)
> XOR<- c(0,1,1,0)
> xor.data <- data.frame(X1,X2,XOR)
> xor.data
```

8.6.2 Modeling Exclusive-OR with MLP: Numeric Output

For our first experiment, let's create a backpropagation neural network with numeric output using MLP. MLP incorporates the previously described sigmoid function for node evaluation. MLP processes both numeric and nominal (categorical) input attributes. In the case of a categorical attribute, the network creates a separate input node for each possible input value. Output attributes can also be either numeric or categorical. If an output attribute is numeric, the network requires but one output node. However, if the output attribute is categorical, each value of the output attribute is assigned an output node. Once the network has been built, an unknown instance is classified with the class whose corresponding output node shows the largest numeric value. Here is the procedure,

Step 1: Identify the Goal

The goal for the XOR problem is straightforward. Given two input values where each value is either a 1 or a 0, we want our network to output the value of the XOR function.

Step 2: Prepare the Data

Scripts 8.2 and 8.3 are found in a single file of your supplementary materials. Load the file containing the scripts into your source editor.

Step 3: Define the Network Architecture

To help define the network architecture, we need to investigate MLPs available options. You can check out the options with these statements:

```
> library(RWeka)
> WOW(MLP)
```

Your output will list all of MLPs' options. Scroll through the list to get a general idea about the options we have at our disposal. Four options are of particular interest. Let's examine each option in greater detail.

– The G option (*GUI parameter*) when set to TRUE allows us to visualize the network prior to and during the training process. The graphical user interface (GUI) lets us control the L (learning rate), M (momentum), and N (Epochs) options. The default value for G is FALSE for good reason. If for example, with a 10-fold cross validation and a setting of true for the GUI parameter, each fold will present us with a newly created network whereby we must respond with a click on accept to continue. Also, it is important to note that when the GUI parameter is true, we can terminate network training or testing prior to completion by simply clicking accept. For our experiment, we wish to see the structure of the created network(s), so let's set the GUI parameter to TRUE.

– The H option (*hiddenLayers* parameter) allows us to specify the total number of nodes in each hidden layer and the total number of hidden layers. The value

4 for this parameter gives us one hidden layer containing four nodes. $H = ('4','3')$ gives us two hidden layers, the first hidden layer having four nodes and the second hidden layer showing three nodes. Deciding on the number of hidden layers and the number of nodes within each hidden layer is both an art and a science. An often used general rule is to divide the sum of the number of classes and the number of attributes by 2. For our first example, give this field a value of 2 to specify one hidden layer with two nodes.

- The B option (*nominalToBinaryFilter*) creates a separate input node for each possible nominal input value. This is designed to improve performance when nominal data are present. To illustrate, if our initial attribute is *income* with nominal values *high*, *middle*, and *low*, the filter will create one input node for each of the three values. That is, the income attribute will actually generate three input nodes. If an input instance has a value of high for the income attribute, the input node representing high within the neural network will receive a value of 1. The nodes representing middle and low income receive zero values.

- The S option (the *seed* parameter) offers us the opportunity to vary the initial values for the network weights. The initial weight assignments can affect the resultant accuracy of the network. Let's leave S at its default value.

Step 4: Watch the Network Train

Use *run* to perform a line-by-line execution of Scripts 8.2 and 8.3. The first line of Script 8.3 gives the call to MLP. The control settings tell us the GUI is on and the network will have one hidden layer of two nodes. Once you invoke MLP, the graphical representation presented in Figure 8.6 defining the network architecture

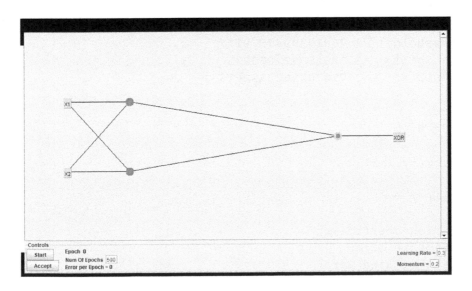

FIGURE 8.6   Architecture for the XOR function.

will be displayed. Figure 8.6 shows that the current parameter setting for Epoch and Error per Epoch are both zero. That is, although the network architecture is displayed, network training has yet to begin. Click *start* to watch the network train. As our training set contains but four instances, training quickly terminates. Notice the value for Epoch is now 500 indicating that the training cycle is complete. In addition, the Error per Epoch is 0 indicating that the network has learned the XOR function. To continue, click *accept* whereby the GUI disappears.

Script 8.3 Learning the XOR Function

```
># The call to MLP

> my.xor <- MLP(XOR ~ ., data = xor.data, control
=Weka_control(G=TRUE,H=2))

> # my.xor and summary tell us about the network
> my.xor

Linear Node 0
    Inputs      Weights
    Threshold         1.201946073787422
    Node 1           -3.1277491093719436
    Node 2            3.2211187739315217
Sigmoid Node 1
    Inputs      Weights
    Threshold         1.0017904224497758
    Attrib X1        -1.8668746389034567
    Attrib X2         1.8070925645762244
Sigmoid Node 2
    Inputs      Weights
    Threshold        -3.7081939947579117
    Attrib X1        -3.185559927055833
    Attrib X2         2.7127022622592616
Class
    Input
    Node 0

> summary(my.xor)

=== Summary ===

Correlation coefficient                   1
Mean absolute error                       0
Root mean squared error                   0
Relative absolute error                   0       %
Root relative squared error               0       %
```

```
Total Number of Instances                    4

> #The training data is also used for testing.
> my.pred <-round(predict(my.xor,xor.data),3)

> # cbind gives us a way to compare actual and predicted outcome.
> my.PTab<-cbind(xor.data,my.pred)
> my.PTab

   X1 X2 XOR my.pred
    0  0   0        0
    1  0   1        1
    0  1   1        1
    1  1   0        0
```

Step 5: Read and Interpret Summary Results

Continuing with Script 8.3, *my.xor* displays the trained network connection weights. Interpreting this output can be very confusing. To help you understand how the weights are represented in the network, we created Figure 8.7, which gives the rounded connection weights as they would appear on the trained network. The listing first shows *Node 0* which represents the lone output node. The *node 1* and *node 2* values listed under the heading *Linear Node 0* represent the *node 1* to

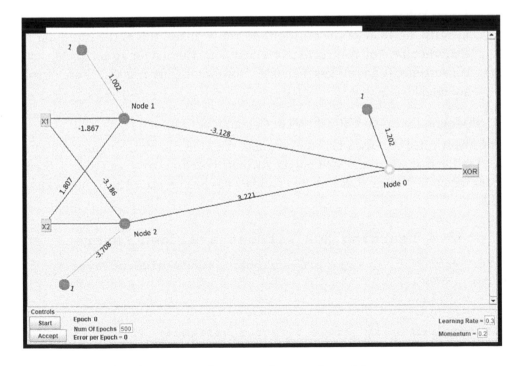

FIGURE 8.7   Network architecture with associated connection weights.

node 0 and *node 2* to *node 0* connection weights. Similarly under *Sigmoid Node 1*, we see the respective connection weights between *X1* and *node 1* as well as *X2* and *node 1*. The same scheme applies for the values under *Sigmoid Node 2*.

Figure 8.7 also shows an additional weighted link associated with each node. The weights are the rounded *threshold* values listed in Script 8.3. The threshold serves a purpose similar to the constant term in a regression equation and functions as another connection weight between two nodes. The 1 at one end of the link represents a node whose output is always 1. The threshold values change during the learning process in the same manner as all other network weights.

In the simplest terms, the *threshold* or *bias* can be thought of as a value added to the associated node. To see this, consider Figure 8.7 and *Node 1* having *threshold* = 1.002 together with the instance showing *X1* = 1 and *X2* = 0. The rounded activation value given to *Node 1* for processing by the sigmoid function will be (–1.867 * 1 + 1.807 * 0 + 1.002) = –0.874. Similar reasoning pertains to the threshold values associated with *Node 0* and *Node 2*.

Returning to Script 8.3, the summary statement shows zero values for the mean absolute error and the root mean squared error. To lend support to these values, we use the *predict* function. Notice how *round* and *cbind* are employed to give us a table of actual and predicted values. Removing the round function will print the output to more closely reflect the exact values determined by the network.

As the training data represents all possible values for the XOR function, we can be assured that—given correct input values—our network will always correctly identify the value of the function. As you will see, this is the exception rather than the rule. You may recall that a common criterion for an acceptable network convergence is an *rms* less than 0.10. However, higher *rms* values are oftentimes acceptable.

8.6.3 Modeling Exclusive-OR with MLP: Categorical Output

Having investigated the case where the output of the XOR function is numeric, it's time to proceed to the case where the XOR output attribute is defined to be categorical. Script 8.4 provides the code. Notice XOR with its *1's* and *0's* has been replaced by XORC with values *a* and *b*.

Script 8.4 Learning XOR without a Hidden Layer

```
> # XOR with MLP, Categorical Output, and no hidden layer
>library(RWeka)
> X1<- c(0,1,0,1)
> X2<- c(0,0,1,1)
# MLP requires output to be a factor variable
XORC<- as.factor(c("a","b","b","a"))
> xorc.data <- data.frame(X1,X2,XORC)
```

```
> xorc.data
  X1 X2 XORC
  0  0   a
  1  0   b
  0  1   b
  1  1   a

> # The call to MLP
> my.xorc <-MLP(XORC ~ ., data = xorc.data,
control=Weka_control(G=TRUE,H=0))

> # A call to summary provides the output.
> summary(my.xorc)

=== Summary ===

Correctly Classified Instances          2          50     %
Incorrectly Classified Instances        2          50     %
Kappa statistic                         0
Mean absolute error                     0.5
Root mean squared error                 0.5001
Relative absolute error                 99.9999 %
Root relative squared error             100.021  %
Total Number of Instances               4

=== Confusion Matrix ===

  a b    <-- classified as
  0 2 |  a = a
  0 2 |  b = b
```

For the first experiment, let's verify that MLP without a hidden layer cannot learn the XOR function. This is accomplished by setting H to 0 prior to invoking the call to MLP. Figure 8.8 provides the network configuration resulting from the call to MLP. Let's once again use the default settings for all parameters. Click *start* to begin training. Notice that after 500 epochs, the network still shows an error value greater than 0.25. Click *accept*.

The summary function in Script 8.4 displays error rate information as well as a confusion matrix resulting from the training data. It is clear that all four instances have been classified as belonging to the class whose XOR output is *b*. This gives us a classification correctness of only 50%. Also, a call to *predict(my.xorc, xorc.data)*—not shown—will display *b b b b*.

Let's try to do better by incorporating more epochs. Invoke MLP a second time. Before clicking *start*, set the value inside the *Num of Epochs* box to 50,000—don't be concerned if the change in value isn't readily visible. You will again see the network converges at a value

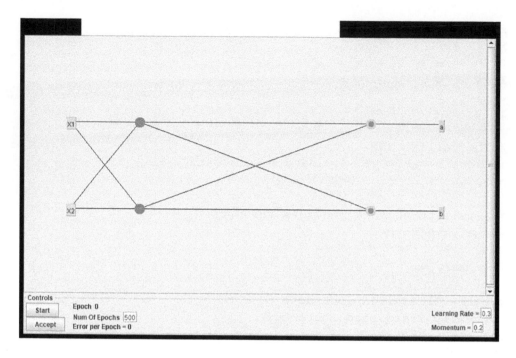

FIGURE 8.8   XOR network architecture without a hidden layer.

greater than 0.25. After a few experiments, it becomes clear that any changes you make other than adding a hiding layer will be futile!

In order to achieve a positive result with XORC, we return to step 3 of our five-step process and change the value of *H*. Changing the value to *1* or *2* while using 50,000 epochs or more reduces the error rate and results in three of four correct classifications. However, you will discover that a hidden layer of at least three with a value for *Num of Epochs* somewhere around 2000 will give a classification correctness of 100%.

8.6.4 Modeling Exclusive-OR with neuralnet: Numeric Output

Before turning to a new dataset, the *neuralnet* function found within the *neuralnet package* is worth our attention. This function is of interest as it allows us greater control over the learning environment when compared to MLP. However, more freedom comes at a price! With MLP, preprocessing matters such as data normalization, categorical to numeric conversions, initializing the randomization seed, and missing data issues are handled internally. This is no longer the case. Also, the job of creating a confusion matrix for categorical output data now rests on our shoulders.

Script 8.5 offers the code for using *neuralnet* to model XOR when the output is numeric. You can see the list of arguments and returned values by typing *help(neuralnet)*. Here we look at the arguments of most interest. You can certainly import the code from the supplementary materials but creating Script 8.5 yourself will give you a better learning experience. That said, a line-by-line analysis of Script 8.5 is in order.

The first lines of Script 8.5 define the data frame for the XOR function. As the weights of the network are initialized with random values, the *set.seed* function is necessary for a consistent result. Next is the call to *neuralnet* where we have set the number of hidden layer nodes to 3. The error and activation functions are set at their default values. *sse* computes overall root mean square error and *logistic* represents the sigmoid function. The *linear. output* parameter is set at FALSE. This tells us that the output nodes will apply the logistic function to their input. Linear.output should be set at TRUE provided there is a linear relationship between the input and output attributes.

summary(my.nnet) has been commented out. However, invoking the summary statement will provide you with a list of several summary parameters. Three are of interest here.

- my.nnet$net.result gives the final output of the neural network for the training data.

- my.nnet$result.matrix displays several values including the error rate and the final network connection weights.

- Mynnet$act.fct outputs the code for the activation function.

The *plot* function gives us Figure 8.9. Notice that it took 97 passes of the data through the network for it to converge to an error rate less than 0.02. As an experiment, change the hidden layer value to '0'. You will see an immediate network convergence with an error rate around 0.50 which is equivalent to guessing as the output value of the network.

The last lines of Script 8.5 show how the predict function is used to test the network with the training data. The cbind and round functions are used to display the input and output in a tabular format. In the next section, we use *neuralnet* to model the categorical output version of XOR.

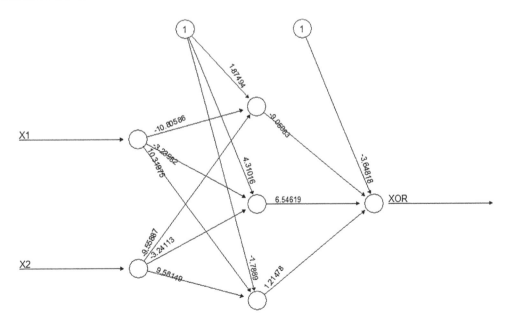

Error: 0.017206 Steps: 97

FIGURE 8.9 Modeling XOR with the neuralnet function.

Script 8.5 Modeling XOR with the neuralnet Function

```
> # Define the XOR function
> X1<- c(0,1,0,1)
> X2<- c(0,0,1,1)
> XOR<- c(0,1,1,0)
> xor.data <- data.frame(X1,X2,XOR)
> #Load the neuralnet package and set the seed
> library(neuralnet)
> set.seed(1000)
> my.nnet <- neuralnet(XOR ~ .,data=xor.data,hidden=3,err.fct = "sse"
+act.fct = 'logistic', linear.output = FALSE)
> ## summary(my.nnet)
> plot(my.nnet)


> #The training data is also used for testing.
> my.pred <-round(predict(my.nnet,xor.data),3)
> # Use cbind to make a list of actual and predicted output.
> my.result<-cbind(xor.data,my.pred)
> my.result

  X1 X2 XOR my.pred
   0  0   0   0.008
   1  0   1   0.920
   0  1   1   0.919
   1  1   0   0.146
```

8.6.5 Modeling Exclusive-OR with neuralnet: Categorical Output

Modeling the XOR function with *neuralnet* when the output is categorical requires more work. Script 8.6 details the approach. Several items require explanation. We first see the 1's and 0's for XORC are now "a's" and "b's". The call to *neuralnet* shows a hidden layer of three nodes.

The graph resulting from the call is given in Figure 8.10. An error rate less than 0.03 indicates a positive result.

Next, the predicted values are stored in *my.pred* which is then structured as a data frame. As there are two classes, the data frame—*my.results*—is a two-column structure where each row shows the numeric value of the corresponding output node. The column labeled *Predicted.1* represents class *a* and the column *Predicted.2* shows class *b* probability values. To see this, consider the output of the *str* function as applied to xorc.data. For XORC, we have,

```
XORC: Factor w/ 2 levels "a","b": 1 2 2 1
```

Recall that factors are stored as integer values with the levels listed by default in alphabetical order.

To make the confusion matrix, we use the *ifelse* structure together with *Predicted.1*—the column representing *a*—to create a list of 1s and 2s. The *factor* function converts the 1s and 2s to their corresponding factor levels. The factor levels are then used by the *table* function to construct the confusion matrix.

Finally, confusionP is applied to the confusion matrix to give an accuracy level for the predictions. Now that we have the basics of MLP and *neuralnet*, let's look at a more interesting problem.

Script 8.6 Modeling XORC with the neuralnet Function

```
> # Define XORC function

> X1<- c(0,1,0,1)
> X2<- c(0,0,1,1)
> XORC <- c("a","b","b","a")
> xorc.data <- data.frame(X1,X2,XORC)

> #Load the neuralnet package
> library(neuralnet)
> # Use set.seed for a consistent result
> set.seed(1000)

# Build and graph the network model.
> my.nnetc <- neuralnet(XORC ~ .,
+             data=xorc.data,hidden=3,
+             act.fct = 'logistic', linear.output = FALSE)
> plot(my.nnetc)

# my.pred represents the network's predicted values
> my.pred<- predict(my.nnetc,xorc.data)

> # Make my.results a data frame
> my.results <-data.frame(Predicted = my.pred)
> my.results
   Predicted.1 Predicted.2
1  0.99498423  0.00335269
2  0.06280405  0.89770257
3  0.05990612  0.90152348
4  0.89814783  0.12229582

> # Use an ifelse statement to convert probabilities to
> # factor values. Column 1 represents "a"
> # Use column 1 for the ifelse statement.
> my.predList<- ifelse(my.results[1] > 0.5,1,2) #  > .5 is an "a"

> my.predList <- factor(my.predList,labels=c("a","b"))
```

```
> my.predList
[1] a b b a
Levels: a b
> my.conf <- table(xorc.data$XORC,my.predList,dnn=c("Actual","Pred
icted"))
> my.conf
      Predicted
Actual a b
     a 2 0
     b 0 2

Correct= 4 Incorrect = 0
Accuracy = 100 %
```

8.6.6 Classifying Satellite Image Data

The satellite image dataset represents a digitized satellite image of a portion of the earth's surface. The training and test data consist of 300 pixels for which *ground truth* has been established. Ground truth of a satellite image is established by having a person on the ground measure the same thing the satellite is trying to measure (at the same time). The answers are then compared to help evaluate how well the satellite instrument is doing its job.

These data have been classified into 15 categories: Urban, Agriculture 1, Agriculture 2, Turf/Grass, Southern Deciduous, Northern Deciduous, Coniferous, Shallow Water, Deep Water, Marsh, Shrub Swamp, Wooded Swamp, Dark Barren, Barren 1, and Barren 2. Each

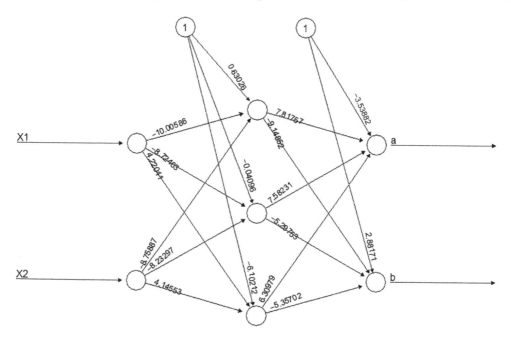

Error: 0.026531   Steps: 89

FIGURE 8.10   Modeling XORC with the neuralnet function.

category contains approximately 20 instances. Each pixel is represented by six numeric values consisting of the multispectral reflectance values in six bands of the electromagnetic spectrum: blue (0.45–0.52 m), green (0.52–0.60 m), red (0.63–0.69 m), near infrared (0.76–0.90 m), and two middle infrared (1.55–1.75 and 2.08–2.35 m). The input data is strictly numeric.

Step 1: Identify the Goal

Script 8.7 shows the steps used to model the data. Our goal is to build a neural network that can be used to monitor land cover changes in the region defined by the dataset. Once a network architecture is determined to be acceptable, the model can be used to monitor for significant changes in the specified region. We will accept a model that shows a test set accuracy greater than or equal to 95%. Upon achieving the desired accuracy, the entire dataset will be used to build the final model.

Step 2: Prepare the Data

Import the .csv form of *Sonar* into RStudio. With the exception of turf grass with 21 instances and deep water with 19 instances, all classes have 20 instances evenly split between the first and second half of the dataset. This makes a case for using half of the data for training and the remaining half for testing. This is shown in Script 8.7 where preprocessing scales the data and defines the training and test datasets.

When building neural network models, it is instructive to understand the importance of attribute selection. This is the case as, unlike models such as decision trees where attribute selection is part of the modeling algorithm, neural network algorithms are unable to determine a priori the importance of any given input attribute. Therefore, it is our job to use the preprocessing phase to make appropriate input attribute selections. In our first examples, attribute preprocessing was trivial as XOR is a well-defined function where all instances and attributes are relevant. However, with the satellite image data, this is not the case. On the other hand, neural network algorithms are quite resilient in that they are oftentimes able to build useful models even when attribute selection is not optimal. End-of-chapter Exercise 7 asks you to experiment with this dataset to determine if any of the input attributes should be eliminated. For our example, we will initially use all six input attributes.

Step 3: Define the Network Architecture

We will employ the *neuralnet* function to define the network for modeling the data. Notice the hidden layer parameter is set at 5. Given the total number of classes, we would normally use a much larger hidden layer. Previous tests with this dataset (Roiger, 2016) have shown that it is very well defined.

Step 4: Watch the Network Train

Figure 8.11 displays the created network. The *rms* is not shown in the figure, but we can obtain it from the *result.matrix*. That is,

```
> my.nnet$result.matrix
                            [,1]
error                  2.728610e-02
reached.threshold      9.992989e-03
steps                  1.575000e+03
```

This tells us that the network converged with an *rms* < 0.03.

Step 5: Read and Interpret Summary Results

To create the confusion matrix, we need a function that—for each test instance— determines the winning node and stores the numeric factor matching the node in a list of predicted values. Each value in this list will be a positive integer ranging between 1 and 15. The function *pred.subs* written specifically for this task does the job by placing the values in *my.predList*. The list of actual classes and *my.predList* is used by the *table* function to create the confusion matrix. The source code for *pred.subs* can be found in *functions.zip*.

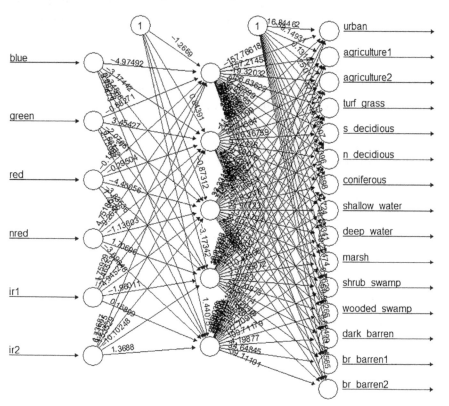

FIGURE 8.11   Modeling satellite image data.

Continuing with Script 8.7, we see a test set accuracy of 94.67% which is extremely close to our desired result. We can do even better by using a single hidden layer of ten rather than five nodes. Doing so results in a 96% test set correctness with all but five test set instances correctly classified.

Given that we are satisfied that we have achieved our goal of creating a good model for classifying land cover, the next step is to use all of the data to create and save a final model. This model will be used in the future to classify the same regions after specified time periods in order to determine if the land cover of any of the regions has changed.

Lastly, our preprocessing did not consider possible correlations among the input attributes. Checking for high correlational values between the input attributes can be accomplished by first removing the output attribute and applying the *cor* function as follows:

```
>Sonar2 <- Sonar[-7]
>cor(Sonar2)
```

This result gives us correlational values between red, green, and blue above 0.85. This indicates that eliminating two of the three attributes should not negatively affect test set accuracy. Here's an easy way to remove green and red from the data.

```
>Sonar2<- Sonar[,c(-2,-3)]
```

We leave it as an exercise to determine the effects of removing these two input attributes.

Script 8.7 Classifying Satellite Image Data

```
> library(neuralnet)
> # Use set.seed for a consistent result
> set.seed(1000)
> Sonar2<- Sonar

> # scale the data
> Sonar2 <- scale(Sonar2[-7])
> # Add back the outcome variable
> Sonar2 <- cbind(Sonar[7],Sonar2)

> # call neuralnet to create a single hidden layer of 5 nodes.
> Sonar2.train <- Sonar2[1:150, ]
> Sonar2.test <- Sonar2[151:300, ]
> my.nnet <- neuralnet(class ~ ., data=Sonar2.train,
+ hidden=5,act.fct = 'logistic', linear.output = F)
> plot(my.nnet)
```

```
> # my.pred represents the network's predicted values
> my.pred<- predict(my.nnet,Sonar2.test)

> # Place the results in a data frame.
> my.results <-data.frame(Predicted=my.pred)

> # Call pred.subs to create the table needed to make
> # the confusion matrix.
> my.predList <- pred.subs(my.results)

> # Structure the confusion matrix
> my.conf <- table(Sonar2.test$class,my.predList)
> # Add a column of matching class numbers
> cbind(1:15,my.conf)
```

		1	2	3	4	5	6	7	8	9	10	11	12	13	14	15
agriculture1	1	10	0	0	0	0	0	0	0	0	0	0	0	0	0	0
agriculture2	2	0	10	0	0	0	0	0	0	0	0	0	0	0	0	0
br_barren1	3	1	0	6	0	0	3	0	0	0	0	0	0	0	0	0
br_barren2	4	0	0	0	11	0	0	0	0	0	0	0	0	0	0	0
coniferous	5	0	0	0	0	10	0	0	0	0	0	0	0	0	0	0
dark_barren	6	0	0	0	0	0	10	0	0	0	0	0	0	0	0	0
deep_water	7	0	0	0	0	0	0	8	0	0	0	1	0	0	0	0
marsh	8	0	0	0	0	0	0	0	8	0	0	0	0	0	0	2
n_decidious	9	0	0	1	0	0	0	0	0	9	0	0	0	0	0	0
s_decidious	10	0	0	0	0	0	0	0	0	0	10	0	0	0	0	0
shallow_water	11	0	0	0	0	0	0	0	0	0	0	10	0	0	0	0
shrub_swamp	12	0	0	0	0	0	0	0	0	0	0	0	10	0	0	0
turf_grass	13	0	0	0	0	0	0	0	0	0	0	0	0	10	0	0
urban	14	0	0	0	0	0	0	0	0	0	0	0	0	0	10	0
wooded_swamp	15	0	0	0	0	0	0	0	0	0	0	0	0	0	0	10

```
Correct= 142 Incorrect= 8
Accuracy = 94.67 %
```

8.6.7 Testing for Diabetes

The dataset—*Diabetes*—contains information about 768 females, 268 of whom tested positive for diabetes. The data include eight numeric input attributes and a categorical output attribute indicating the outcome of a test for diabetes. The dataset is part of the supplementary materials and comes in both .csv and MS Excel formats. You can read more about this dataset on the description page of the Excel spreadsheet housing the data. Let's use our five-step approach to build a model of these data.

Step 1: Identify the Goal

We state the goal as a question. Given the input attributes, can we build a model to accurately determine if an individual in the dataset tested positive for diabetes? As a secondary goal, the model should error on the side of reporting false positives rather than false negatives.

In general, a diagnostic model of this type can have value in several situations. If we can determine with a high degree of accuracy whether an individual has a certain disease without a medical test, we may be able to allow the individual to avoid a costly or possibly painful test. Also, the needed test may not be immediately available and a preliminary result obtained from the model may give direction as to an initial treatment plan. Lastly, a nonobvious relationship between several of the input attributes and the output attribute could lead to a better understanding of the disease in question.

Step 2: Prepare the Data

Script 8.8 displays the method used to model the data. You will notice that the data is first scaled then randomized. The training set contains two thirds of the instances. A correlational test between the input attributes did not reveal any correlations greater than 0.55. All input attributes are used to build the initial model.

Step 3: Define the Network Architecture

The *neuralnet* function defines the network. The network is built with a single hidden layer of six nodes.

Step 4: Watch the Network Train

Figure 8.12 displays the trained network. The network converged with an *rms* above 48.

Step 5: Read and Interpret Summary Results

Script 8.8 shows a test set accuracy of 76.17% with 28 false positives and 33 false negatives. These results are less than optimal. We have several alternatives for possible improvement. One obvious approach is to vary the hidden layer settings. As a second choice, we might try a different technique such as ensemble learning. Attribute elimination is another possibility.

An interesting approach is to use unsupervised clustering to help determine how well the input attributes differentiate the two classes. We demonstrate this last approach in the next section.

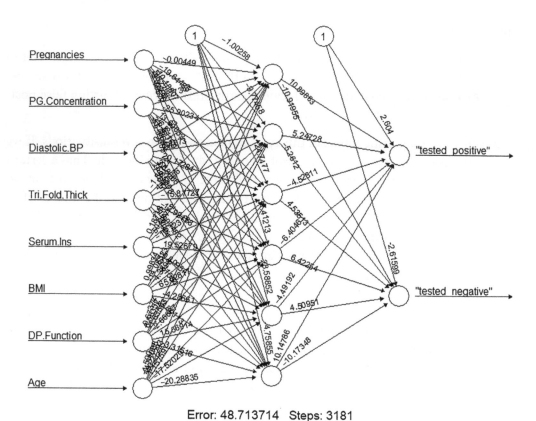

Error: 48.713714 Steps: 3181

FIGURE 8.12   Modeling diabetes data.

Script 8.8 Modeling Diabetes Data

```
> library(neuralnet)
> # scale the data
> sca.dia <- scale(Diabetes[-9])

> # Add back the outcome variable
> sca.dia <- cbind(Diabetes[9],sca.dia)

> # Randomize and split the data for training and testing
> set.seed(1000)
> index <- sample(1:nrow(sca.dia), 2/3*nrow(sca.dia))
> my.Train <- sca.dia[index,]
> my.Test <- sca.dia[-index, ]

> my.nnetc <- neuralnet(Diabetes ~ .,
+              data=my.Train,hidden=6,
+              act.fct = 'logistic', linear.output = FALSE)
> plot(my.nnetc)
```

```
> # Make predictions on the test data.
> my.pred<- predict(my.nnetc,my.Test)

> # Make the table needed to create the confusion matrix.
> my.results <-data.frame(Predictions =my.pred)

> # ifelse converts probabilites to factor values.
> # Use column 1 of my.results for the ifelse statement.
> # Column 1 values are probabilities for tested_negative

> my.predList<- ifelse(my.results[1] > 0.5,1,2) #
>.5="tested_negative"
> # Structure the confusion matrix
> my.predList <- factor(my.predList,labels=c("Neg","Pos"))
> my.conf <- table(my.Test$Diabetes,my.predList,dnn=c("Actual",
"Predicted"))
> my.conf
                         Predicted
Actual                   Neg Pos
   "tested_negative" 153   28
   "tested_positive"  33   42

> # Output accuracy
> confusionP(my.conf)


   Correct= 195 Incorrect= 61
   Accuracy = 76.17 %
```

8.7 NEURAL NET CLUSTERING FOR ATTRIBUTE EVALUATION

Script 8.1 showed you how to install RWeka's implementation of the KSOM. Recall that we associated the variable KSOM with our implementation. Here we use a feature of KSOM together with the diabetes dataset described in the previous section that allows us to compare the clusters formed by a Kohonen network to the actual classes within the data. Here is the approach:

Step 1: Identify the Goal

We will use unsupervised clustering to determine how well the input attributes of the Diabetes dataset differentiate between the actual classes in the data. Clusters that match closely with the classes lend support to the ability of the input attributes to structure a predictive model. If the formed clusters do not match with the classes, we will conclude that the input attributes may not accurately define the structure of the classes.

Step 2: Prepare the Data

Script 8.9 describes the clustering procedure used to analyze how well the input attributes of the Diabetes dataset define the outcome attribute. Prior to calling KSOM, we must remove the class attribute. There is no need to normalize the data as KSOM does this for us.

Step 3: Define the Network Architecture

You can see the control options available with KSOM by typing WOW(KSOM). The options of immediate interest are H and W. H sets the height and W sets the width of the network structure. As there are two classes in the data, we set the height at 2 and the width at 1.

Step 4: Watch the Network Train

The entire dataset is used to cluster the data. When the clustering is complete, each instance will have been assigned to either cluster 0 or 1. The variable $class\_ids$ gives us the cluster number associated with each instance. The following statements place the list of cluster assignments in the variable *ids* and convert *ids* to a single-column 768 row data frame of 0s and 1s.

```
>ids <- my.cluster$class_ids
>ids <- data.frame(ids)
```

In **order** to create the confusion matrix, the factors of the output attribute (1 and 2) must be matched with the cluster numbers (0 and 1). *Tested-negative* is associated with factor 1 and *tested-positive* with factor 2 as *tested-negative* is alphabetically first. If cluster 0 matches *tested-negative*, we must convert all 0's in *ids* to 1's and all 1's to 2's. How do we know?

With clustering, there is no output attribute so *cluster 0* is simply associated with the first instance in the dataset. In our case, unless we look at the assigned class of the first instance, we don't know if cluster 0 matches with *tested-negative* or *tested-positive*. Another problem arises if the first instance is an outlier and ends up clustering with the opposite class. Fortunately, a little domain knowledge comes to the rescue! Upon examination of the mean and standard deviation values in Script 8.9, we see that the average age of the individuals seen in cluster 1 is much higher than the average age shown in cluster 0. As diabetes is more prevalent as we age, we associate cluster 1 with the *tested_positive* class. The for loop in the script reflects this observation and changes all 1s in *ids* to 2s and 0s to 1s. When domain knowledge is completely lacking, we initially modify the *ids* in this same way. If the resultant confusion matrix looks "backwards", we must reassess and possibly modify the loop by reassigning all 0s as 2s and all 1s as 0s. With three or more clusters, similar reasoning applies.

Step 5: Read and Interpret Summary Results

An abbreviated listing of the contents of *my.cluster* tells us that between-cluster means differ for attributes *age* and *pregnancies*. The remaining input attributes show only small between-cluster differences. The confusion matrix displays 138 false negative and 118 false positive classifications. This tells us that the clusters do not match well with the actual classes. This, in part, explains the poor test set results in our previous experiment. End-of-chapter exercise 8 asks you to further investigate the relevance of these input attributes.

Script 8.9 Clustering Diabetes Data

```
> library(RWeka)
> WPM("load-packages",KSOM)

> # Cluster the Diabetes Data
> # We must first remove the output attribute.
> # As we want two classes, we set H=2 and W=1
> diabetes.data <- Diabetes[-9]

> # Cluster the data
> my.cluster <- KSOM(diabetes.data,control =
Weka_control(H=2,W=1))
> my.cluster

Self Organized Map
==================
Number of clusters: 2
```

	Cluster	
Attribute	0	1
	(520)	(248)
Pregnancies		
mean	2.0981	7.5081
std. dev.	1.7667	2.9667
PG.Concentration		
mean	115.5192	132.1653
std. dev.	30.585	31.9439
Diastolic.BP		
mean	66.1519	75.2984
std. dev.	19.771	16.8866
Tri.Fold.Thick		
mean	21.8231	17.8387
std. dev.	15.0184	17.478
Serum.Ins		
mean	84.3712	70.2137
std. dev.	110.5879	124.1295

```
BMI
  mean                    31.8046   32.3867
  std. dev.                8.2168    7.1364
DP.Function
  mean                     0.4695    0.4769
  std. dev.                0.3408    0.3112
Age
  mean                    26.8865   46.5645
  std. dev.                5.619     9.9941
```

```
> # Associate a cluster number with each instance
>ids <- my.cluster$class_ids

> # create a data frame of one column
> ids <- data.frame(ids)
> ids

# See the explanation given in your text about how to, in
# general set up the for loop below to give
# appropriate values to ids[i,1].

# The following for loop changes ids with a value of 1 to 2
# and ids with a value of 0 to 1 which for these data is
# the correct assignment.

> for (i in 1:nrow(ids))
+ {     if(ids[i,1] == 1)
+          ids[i,1] <- 2
+        else
+            ids[i,1] <-1
+ }
> #ids

> # Print the confusion matrix
> my.conf <- table(Diabetes$Diabetes,ids[,1],dnn=c("Actual",
+ "Predicted"))
> my.conf
                    Predicted
Actual                 0    1
  "tested_negative" 382  118
  "tested_positive" 138  130

> confusionP(my.conf)
  Correct= 512 Incorrect= 256
  Accuracy = 66.67 %
```

8.8 TIMES SERIES ANALYSIS

Oftentimes, the data we wish to analyze contains a time dimension. Prediction applications with one or more time-dependent attributes are called *time series problems*. Time series analysis usually involves predicting numeric outcomes, such as the future price of an individual stock or the closing price of a stock index. Three additional applications suitable for time series analysis include

- Tracking individual customers over time to determine if they are likely to terminate the use of their credit card

- Predicting the likelihood of automotive engine failure

- Predicting the weekly rushing yards of a National Football League running back

Much of the work with time-dependent data analysis has been statistical and limited to predicting the future value of a single variable. However, we can use both statistical and nonstatistical machine learning tools for time series analysis on one or several variables. Fortunately, we are able to apply the same techniques we use for other machine learning applications to solve time series problems. Our ability to succeed is determined to a large extent by the availability of relevant attributes and instances, as well as by the difficulty of the problem at hand. Let's look at an interesting application of time series analysis.

8.8.1 Stock Market Analytics

Human experts exist in fields such as medicine, automotive repair, computer programming, and judicial law to name a few. Although a solid education is part of the process of becoming an expert, experience most often plays the larger role. Unfortunately, all of the experience in the world does not create an "expert stock market analyst". The problem is not only in the large number of variables affecting stock market trends but also in the fact that a change in the value of a variable might be interpreted in several ways. In one case, a rise in interest rates causes stocks to tumble as it now costs companies more to borrow money. In another case, that same rise in rates sends stocks soaring as it means that the economy is improving. One thing is certain, there is no lack of "experts" willing to give a detailed expos facto analyses of any market moving situation! Knowing this, many market analysts work very hard at their trade and tend to perform well most of the time. These individuals fall into one of four general groups.

One set concentrates solely on analyzing company fundamentals. They examine company financial statements and talk with company representatives in an attempt to better understand the company's future. A second group makes their trading choices based solely on recent market trends and trading movements. These individuals are known as technical analysts. These market technicians use stock charts, moving averages, technical support levels, and standard deviations to determine what the future holds for a particular stock or index fund. The third set of individuals combine both techniques relying at times more on fundamentals and at other times more on technical analysis. Then there's the fourth group known as contrarians who base their decisions on market sentiment. These individuals rely on the fact that the market tends to be either too optimistic or too pessimistic at any given time. The contrarian bets against market sentiment, buying when most are selling or selling when the majority of folks are buying. Our interest lies with the group of technical analysts many of whom incorporate time series analysis through machine learning as part of their technical toolkit. Let's see how they do it!

8.8.2 Time Series Analysis: An Example

For our example, we use an exchange traded fund (ETF) with symbol SPY. An ETF is an index fund that can be bought or sold just like a stock. For example, a commodity index fund such as GLD (gold) or USO (United States oil) trades in unison with the corresponding commodity. VGT is a technology index fund where Apple Computer, Microsoft Corporation, Facebook, and Google make up 35% of its holdings. XLF is a financial ETF with top holdings of Berkshire Hathaway, Wells Fargo, and JP Morgan Chase & Company. Owning an index fund is in general a much safer bet than owning individuals stocks as bad news for one stock doesn't necessarily mean disaster for the entire basket of stocks held by the fund. Because of this, the past few years has seen the birth of thousands of ETFs.

The fund of interest to us is SPY, a fund that mimics the movement of the Standard and Poor's 500 index. SPY is of special interest to market traders and especially market timers for at least two reasons. First, SPY's value tends to increase over time meaning that if you hold the fund for long periods, you will likely make money or in the worst case, you won't lose your initial investment. More importantly, SPY experiences price swings to both the up and downside making it a frequent market trader's dream. It is not unusual to see SPY trade two or more percentage points higher (or lower) in a single day! If we can build a model to identify when and in what direction SPY moves, the potential for significant profit is excellent!

To begin our experiment with SPY, we start with a goal.

> *Build a time series model able to predict the next day closing price for SPY 15 to 30 minutes before current day market closure.*

The goal clearly states that the prediction comes right before the market closes. In this way, we can make and act on our decision prior to market closure. This is of primary importance as waiting to make the trade until the next day market open may be too late. For example, if SPY is going to do well the next trading day, its next-day opening price will likely be higher than the current day close. Waiting to purchase SPY until the next day would therefore limit our opportunity to make a profit.

If our model proves to be satisfactory, we will use it as follows:

- If the model tells us SPY will show a next trading day gain, we purchase the fund.

- If the model tells us SPY is to close lower, we sell our shares.

- If we don't currently own SPY and our model indicates a lower next day close, we might consider purchasing shares of the ProShares Short(S&P 500) index fund SH as its price movement shows an inverse relationship with SPY.

Lastly, if our model is able to provide numeric output, we can base our decision on the size of the predicted move. If the predicted price move is minimal, we may consider simply holding any owned shares. Is it too early to start adding up our profits? Let's find out!

8.8.3 The Target Data

There are a multitude of ways to acquire historical stock data. For our example, we used yahoo together with functions from the quantmod library to obtain the price of SPY for the dates January 1, 2017, through January 2, 2020. Using only the adjusted closing prices, we structured our time series by building a time lag of four directly into the data. That is, any given row of data contains the current day's closing price followed in order by the fund price for the four previous days.

Script 8.10 shows the process used to obtain the data as well as the function that creates the time-lagged instance dataset. The first line displayed by *round(head(my.tsdata),2)* gives the closing price of SPY for January 3 through 9 of 2017. Specifically, the adjusted closing price for January 3, 2017, shows as $211.55. The closing price for January 9, 2017, is seen as $212.70.

Script 8.10 Create the Time Series Data

```
> #OBTAIN VALUES FOR THE SPY
> library(quantmod)
> setSymbolLookup(SPY =list(name="SPY",src='yahoo'))
> getSymbols("SPY", from = "2017-01-01", to = "2020-1-31")
[1] "SPY"

> head(SPY)
           SPY.Open SPY.High SPY.Low SPY.Close SPY.Volume SPY.Adjusted
2017-01-03   225.04   225.83  223.88    225.24   91366500      211.5526
2017-01-04   225.62   226.75  225.61    226.58   78744400      212.8111
2017-01-05   226.27   226.58  225.48    226.40   78379000      212.6420
2017-01-06   226.53   227.75  225.90    227.21   71559900      213.4028
2017-01-09   226.91   227.07  226.42    226.46   46939700      212.6984
2017-01-10   226.48   227.45  226.01    226.46   63771900      212.6984

> spy.data <- SPY[ ,6]

> #FUNCTION TO CREATE THE TIME SERIES
> myts.create <- function(x,C)
+ { # x is the data to use to create the time series
+   # C is the number of columns in the time series
+   ts.df <- NA
```

```
+    ts.df <- data.frame("spy"=ts.df)
+    ts.df[1:C]<- 0
+    # Step 1 create the first row
+    j<-1 # Index for ts.df
+    for(i in c(C:1)) # Cth day is the starting position
+    {
+      ts.df[1,j]<- x[i]
+      j<-j + 1
+    }
+
+    # change column names
+    colnames(ts.df) <- c("spy","spy-1","spy-2","spy-3","spy-4")
+
+    # Step 2 create remaining rows
+    jrow <-2  # keeps track of the row number
+    for(i in ((C + 1):length(x)))
+    {
+      ts.df[jrow,1]<- x[i]    # Today's closing
+      # put in the previous closings
+      for(k in (1:(C-1)))
+        {
+          ts.df[jrow,k+1]<- ts.df[jrow-1,k]
+
+        }
+      jrow <- jrow +1
+    }
+    return (ts.df)
+ }
> # CALL THE FUNCTION TO CREATE THE TIME SERIES
> my.tsdata <- myts.create(spy.data,5)

> round(head(my.tsdata),2)

    spy     spy-1   spy-2   spy-3   spy-4
1 212.70 213.40 212.64 212.81 211.55
2 212.70 212.70 213.40 212.64 212.81
3 213.30 212.70 212.70 213.40 212.64
4 212.76 213.30 212.70 212.70 213.40
5 213.25 212.76 213.30 212.70 212.70
6 212.50 213.25 212.76 213.30 212.70
```

8.8.4 Modeling the Time Series

Script 8.11 displays the statements for modeling the time series. Vector x contains the closing prices for our 770-day period. These values produce the graph of SPY given in Figure 8.13. The function *ma* located in the *forecast* package is used to produce a 3-day moving average to smooth the data points. With a 3-day moving average, each data point is replaced by the average of the data point, the point before and the data point after it.

Script 8.11 Predicting the Next Day Closing Price of SPY

```
> # Script 8.11 Predict the next day closing price of SPY
> library(RWeka)
> library(forecast)
> # PLOT CHART OF SPY FOR THE GIVEN TIME PERIOD
> x <- my.tsdata[,1]
> # Set maximum and minimum values for plot of moving average
> ylim<- c(min(x),max(x))
> plot(ma(x,3),ylim=ylim, main='SPY Jan. 1,2017 - Jan. 31, 2020')
> myts.data <-my.tsdata

> # USE FIRST nrow(my.tsdata)-1 DAYS FOR TRAINING
> myts.train <- myts.data[1:(nrow(my.tsdata)-1),]
> # DAY nrow(my.tsdata) CLOSE IS TO BE PREDICTED
> myts.test <- myts.data[nrow(my.tsdata),]

> #BUILD THE MODEL
> # The call to MLP
> my.tsModel <- MLP(spy ~ ., data = myts.train, control
=Weka_control(G=TRUE,H=10))

> #PREDICT THE OUTCOME
> my.pred <-round(predict(my.tsModel,myts.test),3)
> # cbind gives us a way to compare actual and predicted outcome
> my.PTab<-cbind(my.pred,myts.test)
> round(my.PTab,2)
```

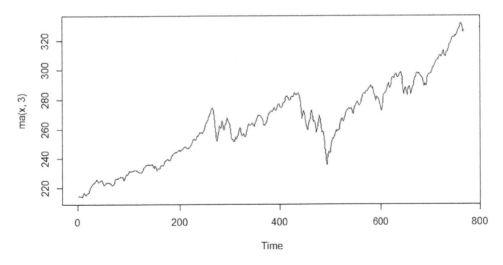

FIGURE 8.13   A 770-day price chart for SPY.

Next, the date column is removed, and the first 769 days are designated for training. We trained the neural network using 5000 epochs. When the test item was presented to the network, it gave a day 770 closing price prediction of $325.60. The actual adjusted close was $325.76 indicating an accurate prediction.

```
       my.pred      spy  spy-1  spy-2  spy-3  spy-4
770    325.6   325.76 324.71 324.98 321.61 326.85
```

8.8.5 General Considerations

This example gives you the general idea about how a neural network can be used to model time series data. However, the approach of using a single variable works well only when the stock market is relatively stable. An improved model will contain more than one time-dependent variable. In any case, it's much too early to start using this model to direct trades in your retirement account! Lastly, although not detailed here, R contains a wealth of very useful functions available for time series analysis.

8.9 CHAPTER SUMMARY

A neural network is a parallel computing system of several interconnected processor nodes. The input to individual network nodes is restricted to numeric values falling in the closed interval range [0, 1]. Because of this, categorical data must be transformed prior to network training.

Developing a neural network involves first training the network to carry out the desired computations and then applying the trained network to solve new problems. During the learning phase, training data is used to modify the connection weights between pairs of nodes so as to obtain a best result for the output node(s).

The feed-forward neural network architecture is commonly used for supervised learning. Feed-forward neural networks contain a set of layered nodes and weighted connections between nodes in adjacent layers.

Feed-forward neural networks are often trained using a backpropagation learning scheme. Backpropagation learning works by making modifications in weight values starting at the output layer then moving backward through the hidden layers of the network.

The self-organizing Kohonen neural network architecture is a popular model for unsupervised clustering. A self-organizing neural network learns by having several output nodes compete for the training instances. For each instance, the output node whose weight vectors most closely match the attribute values of the input instance is the winning node. As a result, the winning node has its associated input weights modified to more closely match the current training instance.

A central issue surrounding neural networks is their inability to explain what has been learned. Despite this, neural networks have been successfully applied to solve problems in both the business and scientific world. Although we have discussed the most popular neural network models, several other architectures and learning rules have been developed. Jain et al. (1996) provide a good starting point for learning more about neural networks.

8.10 KEY TERMS

- *Average member technique.* An unsupervised clustering neural network explanation technique where the most typical member of each cluster is computed by finding the average value for each class attribute.

- *Backpropagation learning. Backpropagation* is a training method used with many feed-forward networks. Backpropagation works by making modifications in weight values starting at the output layer then moving backward through the hidden layer.

- *Bias* or *Threshold Node.* Nodes associated with hidden and output layer. Bias or threshold nodes differ from regular nodes in that their input is a constant value. Corresponding link weights change during the learning process in the same manner as all other network weights.

- *Delta rule.* A *neural* network learning rule designed to minimize the sum of squared errors between computed and target network output.

- *Epoch.* One complete *pass* of the training data through a neural network.

- *Feed-forward neural network.* A neural network architecture where all weights at one layer are directed toward nodes at *the* next network layer. Weights do not cycle back as inputs to previous layers.

- *Fully connected.* A neural network structure where all nodes at one layer of the network are connected to all nodes in the next layer.

- *Kohonen network.* A two-layer *neural* network used for unsupervised clustering.

- *Linearly separable.* Two *classes*, A and B, are said to be linearly separable if a straight line can be drawn to separate the instances of class A from the instances of class B.

- *Neural network.* A parallel *computing* system consisting of several interconnected processors.

- *Neurode.* A neural network processor node. Several neurodes are connected to form a complete neural network structure.

- *Perceptron neural network.* A simple *feed*-forward neural network architecture consisting of an input layer and a single output layer.

- *Sensitivity analysis.* A neural network explanation technique that allows us to determine a rank ordering for the relative importance of individual attributes.

- *Sigmoid function.* One of the several commonly used neural network evaluation functions. The sigmoid function is continuous and outputs a value between 0 and 1.

- *Time series problem.* A *prediction* application with one or more time-dependent attributes.

EXERCISES

Review Questions

1. Draw the nodes and node connections for a fully connected feed-forward network that accepts three input values, has one hidden layer of five nodes, and an output layer containing four nodes.

2. Section 8.1 describes two methods for categorical data conversion. Explain how you would use each method to convert the categorical attribute *income range* with possible values 10–20 K, 20–30 K, 30–40 K, 40–50K … 90–100 K to numeric equivalents. Which method is most appropriate?

3. The average member technique is sometimes used to explain the results of an unsupervised neural network clustering. List the advantages and disadvantages of this approach.

Computational Questions

1. We have trained a neural network to predict the future price of our favorite stock. The 1-year stock price range is a low of $20 and a high of $50.

   a. Use Equation 8.1 to convert the current stock price of $40 to a value between 0 and 1.

   b. Suppose we apply the network model to predict a new price for some time period in the future. The neural network gives an output price value of 0.3. Convert this value to a predicted price we can understand.

2. Consider the feed-forward network in Figure 8.1 with the associated connection weights shown in Table 8.1. Apply the input instance [0.5, 0.2, 1.0] to the feed-forward neural network. Specifically,

   a. Compute the input to nodes i and j.

   b. Use the sigmoid function to compute the initial output of nodes i and j.

   c. Use the output values computed in part b to determine the input and output values for node k.

Experimenting with R

1. Create a backpropagation network using MLP to model one or more of the three logical operators shown in the table below. Because these operators are linearly separable, you might hypothesize that unlike the XOR operator we don't need to use a hidden layer as part of the network architecture. Run experiments to affirm or reject this hypothesis.

2. The complement operator takes a single input value and outputs 1 when the input is 0 and 0 when the input is 1. Use MLP or *neuralnet* to build a backpropagation network to model logical complement. How many epochs are needed to train the network without a hidden layer? If you add a hidden layer, does it take fewer epochs to train the network?

3. Model the categorical version of the XOR function—XORC—with *neuralnet* using

Input	Input 2	And	Or	Implication
1	1	1	1	1
0	1	0	1	1
1	0	0	1	0
0	0	0	0	1

two hidden layers of varying sizes. For example, two hidden layers where the first layer is 3 and the second is 2 is written *hidden = c(3,2)*. Can you find a two hidden-layer network that models XORC without any classification errors?

4. Repeat Exercise 2 but use the *neuralnet* function to build the network.

5. Repeat the experiment with the satellite image data using MLP. A simple approach is to make changes to Script 8.7 as needed. As MLP normalizes data internally, you can begin by defining the training and test datasets. Use the following call to MLP which creates a single hidden layer of five nodes.

    sonar.model<-MLP(class ~ .,data=sonar.train,control=Weka_control(H=5,G=TRUE))

    Use *summary(sonar.model)* to display the confusion matrix created from the training data. Follow this with a call to predict to determine the predictions for the test data. Use the *table* function to print the confusion matrix for the test data. Employ ConfusionP to display test set accuracy. Repeat this experiment but change the number of hidden layer nodes to 4,3,2,1, and finally 0. Display your results in a table. What do you conclude?

6. The *nnet* function within the *nnet package* is another popular function used to develop feed-forward neural networks in R. Install the *nnet package* and read the documentation on how to use the *nnet* function. Use *nnet* to repeat one of the XOR experiments.

7. The satellite image experiment mentioned that a correlational test between the input attributes showed correlations between red, green, and blue above 0.85. Use the *cor* function to verify these correlational values. We know that all but one of a set of highly correlated input attributes should be removed from the data. Remove two of the highly correlated attributes and repeat the satellite image experiment with the modified data. Report your findings. Include the correlation matrix and the resultant confusion matrix in your report.

8. Apply RWeka's InfoGainAttributeEval to the satellite image dataset to determine the effect that attribute elimination has on MLP's classification correctness. Make a table showing the classification accuracy of models built with the best five input attributes followed by the best four attributes down to the best single attribute. Compare your results with those seen when all six input attributes are employed.

9. Repeat Exercise 8 using the *neuralnet* function.

10. This exercise uses the *CardiologyMixed* dataset and consists of two parts.

    a. Use MLP to build a classification model for heart disease. Use the first 200 instances for training and the remaining instances as test data. Experiment with the hidden-layer parameter for a best result. Report the test set results. Provide a detailed explanation of how MLP handles the conversion of categorical data to numeric equivalents.

    b. Use KSOM to cluster the data. Follow the procedure given in Script 8.8. Remove the class attribute and set H at 2 and W at 1 so two classes are formed. KOM always places the first data instance in cluster 0. For this data, the instance is from the sick class. Given this, use the code below for assigning the correct values to ids.

```
for (i in 1:nrow(ids))
{      if(ids[i,1] == 0)
          ids[i,1] = 2
      else
      ids[i,1]=1
 }
```

Do the clusters match well with the defined classes? Base your answer on the confusion matrix created by the clusters-to-classes procedure.

11. Use InfoGainAttributeEval together with the *Spam* datatset to determine the effect attribute elimination has on MLP's classification correctness. Use the first half of the data for training and the remaining instances for testing. Obtain a benchmark by using all input attributes. Next, repeat training and testing with the top ten then the top five "best" attributes. Is there a difference in classification accuracy between each result? Summarize your findings. Be sure to comment on differences between the percentages of false negative and false positive errors seen with each model. Make a general statement about what you have discovered.

12. COVID-19 wreaked havoc with world stock markets during February and March of 2020. To see this, repeat the time series experiment described in Section 8.9 but replace SPY with the Vanguard Total World Stock Index fund VT. Vary the end date of your experiment by choosing 10 alternate dates in March, 2020. Make a table showing how well your model predicted the next day closing price of VT for each chosen date. How well did your model perform in this period of high market volatility?

13. Repeat the time series experiment described in Section 8.9 but replace SPY with the symbol for Facebook (FB), Alibaba (BABA), Amazon (AMZN), or your favorite publically traded company. Report on how your model performs.

Programming Questions

1. Add code to Script 8.5 to show the absolute difference between actual and predicted output for each instance. Your code should also compute and print the mean absolute error and root mean squared error.

2. Create a function that prints a confusion matrix. The function should accept two arguments, the first being the actual class values and the second the predicted values. Place the call to confusionP within the function in order to also report the accuracy of a given model. Here is the skeleton for the function:

```
confusion <- function(actual, predicted)
{      table(......)

       confusionP( ....)

}
```

Installed Packages and Functions

Package Name	Function(s)		
base / stats	c, cbind, data.frame, ifelse, library, max, min, plot, predict round, sample, set.seed, scale, table		
forecast	ma		
quantmod	getSymbols, setSymbolLookup		
neuralnet	neuralnet		
nnet	nnet		
RWeka	GainRatioAttributeEval, OrganizingMap, WOW, make_Weka_clusterer	MultiLayerPerceptron, WPM,	Self make_Weka_classifier,

## CHAPTER 9

Formal Evaluation Techniques

In This Chapter

- Tools for Evaluation

- Confidence Intervals for Model Test Set Accuracy

- Comparing Supervised Models

- Evaluating Supervised Models Having Numeric Output

I N THE PREVIOUS CHAPTERS, we showed how test set error rates, confusion matrices, and receiver operating characteristic (ROC) curves can help us evaluate supervised models. In Chapter 8, you saw how unsupervised clustering can be used for attribute evaluation. In this chapter, we continue our discussion of performance evaluation by focusing on formal evaluation methods for supervised learning. Most of the methods introduced in this chapter are of a statistical nature. The advantage of this approach is that it permits us to associate a level of confidence with the outcome of our experiments.

We emphasize the practical application of standard statistical and nonstatistical methods rather than the theory behind each technique. Our goal is to provide the necessary tools to enable you to develop a clear understanding of which evaluation techniques are appropriate for your data mining applications. The methods presented here are enough to meet the needs of most interested readers. However, Appendix B provides additional material for the reader who desires a more complete treatment of statistical evaluation techniques.

In Section 9.1, we highlight the component parts of the machine learning process that are responsive to an evaluation. In Section 9.2, we provide an overview of several foundational statistical concepts such as mean and variance scores, standard error (SE) computations, data distributions, populations and samples, and hypothesis testing. Section 9.3 offers a method for computing test set confidence intervals for classifier error rates. Section 9.4 shows you how to employ classical hypothesis testing together with test set error rates to compare the classification accuracy of competing models. In Section 9.5, we show you how to evaluate supervised learner models having numerical output. As you read and work through the examples of this chapter, keep in mind that a best evaluation is accomplished by applying a combination of statistical, heuristic, experimental, and human analyses.

9.1 WHAT SHOULD BE EVALUATED?

Figure 9.1 shows the major components used to create and test a supervised learner model. All elements contribute in some way to the performance of a created model. When a model fails to perform as expected, an appropriate strategy is to evaluate the effect every component has on model performance. The individual elements of Figure 9.1 have each been a topic of discussion in one or more of the previous chapters. The following is a list of the components shown in the figure together with additional considerations for evaluation.

1. *Supervised model.* Supervised models are usually evaluated on test data. Special attention may be paid to the cost of different types of misclassification. For example, we might be willing to use a loan application model that rejects borderline individuals who would likely pay off a loan provided the model does not accept strong candidates for loan default. In this chapter, we add to your evaluation toolbox by showing you how to compute test set error rate confidence intervals for supervised models having categorical output.

2. *Training data.* If a supervised model shows a poor test set accuracy, part of the problem may lie with the training data. Models built with training data that does not represent the set of all possible domain instances or contains an abundance of atypical instances are not likely to perform well. A best preventative measure is to randomly select training data, making sure the classes contained in the training data are distributed as they are seen in the general population. The procedure for ensuring an appropriate distribution of data is known as *stratification.*

3. *Attributes.* Attribute evaluation focuses on how well the attributes define the domain instances. We have seen that all but one of a set of highly correlated input attributes should be removed from the data.

4. *Model builder.* It has been shown that supervised models built with alternative learning techniques tend to show comparable test set error rates. However, there may be situations where one learning technique is preferred over another. For example,

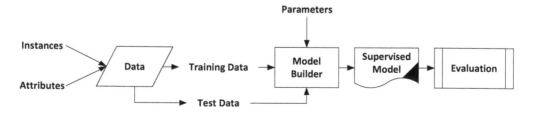

FIGURE 9.1   Components for supervised learning.

neural networks tend to outperform other supervised techniques when the training data contains a wealth of missing or noisy data items. In Section 9.4, we show you how to decide if two supervised learner models built with the same training data show a significant difference in test set performance.

5. *Parameters*. Most data mining models support one or more user-specified learning parameters. Parameter settings can have a marked effect on model performance. The technique used to compare supervised learner models built with different machine learning techniques can also be applied to compare models constructed with the same method but alternate settings for one or more learning parameters.

6. *Test set evaluation*. The purpose of test data is to offer a measure of future model performance. Test data should be selected randomly, with stratification applied as appropriate.

Figure 9.1 also applies to unsupervised clustering, with the exception that we are without test data containing instances of known classification.

9.2 TOOLS FOR EVALUATION

Statistics are a part of our everyday lives. The results of statistical studies help us make decisions about how to invest our hard-earned money, when to retire, where to place our gambling bets, and even whether to have a certain surgical procedure. The following are several interesting statistical findings:

- 60% of the 127,000,000 households in the United States have credit card debt of $6000 or more.

- 50% of all adults in the United States are not invested in the stock market.

- 70% of all felons come from homes without a father.

- One in 700,000 deaths is caused by dog bite.

- The average age when a woman becomes a widow is 55.

- Approximately 70% of financial professionals do not extend their education beyond initial licensing and continuing education requirements.

- One in five adults over 65 have been victimized by financial swindle.

Sometimes, the results of statistical studies must be interpreted with a degree of caution. For example, the average age when a woman becomes a widow may be 55; however, the *median* age when a woman is widowed is likely to be higher as well as more informative. A recent radio commercial for health insurance made the statement "50% of all Americans go bankrupt because of fatal illness." The correct statement is "50% of all Americans who have a fatal illness go bankrupt!"

Findings such as those just listed are often gathered through a process of random sampling. The first statement fits this category. It is simply too difficult to poll each and every American family to determine the amount of household credit card debt. Therefore, experts poll a sample of individual households and report findings in terms of the general population along with a margin of error. As you will see, we can apply the techniques developed to conduct statistical studies to our machine learning problems. The advantage of a statistical approach is that it allows us to associate levels of confidence with the outcome of our experiments.

Before investigating several useful statistics for evaluating and comparing models, we review the fundamental notions of mean, variance, and population distributions.

9.2.1 Single-Valued Summary Statistics

A population of numerical data is uniquely defined by a mean, a standard deviation, and a frequency or probability distribution of values occurring in the data. The *mean*, or average value, denoted by μ is computed by summing the data and dividing the sum by the number of data items.

Whereas the mean designates an average value, the *variance* (σ^2) represents the amount of dispersion about the mean. To calculate the variance, we first compute the sum of squared differences from the mean. This is accomplished by subtracting each data value from the mean, squaring the difference, and adding the result to an accumulating sum. The variance is obtained by dividing the sum of squared differences by the number of data items. The *standard deviation*, denoted by σ, is simply the square root of the variance.

When computing a mean, variance, or standard deviation score for a sampling of data, symbol designations change. We adopt the following notation for sample mean and variance scores:

$$\text{Sample mean} = \overline{X}$$

$$\text{Sample variance} = V$$

The mean and variance are useful statistics for summarizing data. However, two populations can display very similar mean and variance scores yet show a marked variation between their individual data items. Therefore, to allow for a complete understanding of the data, knowledge about the distribution of data items within the population is necessary. With a small amount of data, the data distribution is easily obtainable. However, with large populations, the distribution is often difficult to determine.

9.2.2 The Normal Distribution

A fundamentally important data distribution that is well understood is the *normal distribution,* also known as the Gaussian curve or the normal probability curve. Several useful statistics have been developed for populations showing a normal distribution.

The normal, or bell-shaped, curve was discovered by accident in 1733 by the French mathematician Abraham de Moivre while solving problems for wealthy gamblers.

The discovery came while recording the number of heads and tails during a coin-tossing exercise. For his experiment, he repeatedly tossed a coin ten times and recorded the average number of heads. He found that the average as well as the most frequent number of heads tossed was five. Six and four heads appeared with the same frequency and were the next most frequently occurring numbers. Next, three and seven heads occurred equally often, followed by two and eight and so on. Since Moivre's initial discovery, many phenomenon such as measures of reading ability, height, weight, intelligence quotients, and job satisfaction ratings, to name a few, have been found to be distributed normally. The general formula defining the normal curve for continuous data is uniquely determined by a population mean and standard deviation and can be found in Appendix B.

A graph of the normal curve is displayed in Figure 9.2. The x-axis shows the arithmetic mean at the center in position 0. The integers on either side of the mean indicate the number of standard deviations from the mean. To illustrate, if data is normally distributed, approximately 34.13% of all values will lie between the mean and one standard deviation above the mean. Likewise, 34.13% of all values are seen between the mean and one standard deviation below the mean. That is, we can expect approximately 68.26% of all values to lie within one standard deviation on either side of the mean score.

As an example, suppose the scores on a test are known to be normally distributed with a mean of 70 and a standard deviation of 5. Knowing this, we can expect 68.26% of all students to have a test score somewhere between 65 and 75. Likewise, we should see over 95% of the student scores falling somewhere between 60 and 80. Stated another way, we can be 95% confident that all student test scores lie between two standard deviations above and below the mean score of 70.

As most data is not normally distributed, you may wonder as to the relevance of this discussion to machine learning. After all, even if one attribute is known to be normally distributed, we must deal with instances containing several numeric values, most of which are not likely to be distributed normally. Our discussion of the normal distribution serves two purposes. First, some machine learning techniques assume numeric attributes to be normally distributed. More importantly, as you will see in the remaining sections of this chapter, we can use the properties of the normal distribution to help us evaluate the performance of our machine learning models.

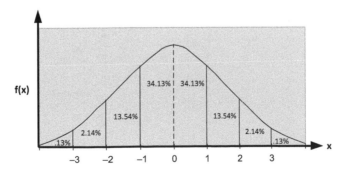

FIGURE 9.2 A normal distribution.

9.2.3 Normal Distributions and Sample Means

Most interesting populations are quite large, making experimental analysis extremely difficult. For this reason, experiments are often performed on subsets of data randomly selected from the total population. Figure 9.3 shows three sample datasets each containing three elements that have been taken from a population of ten data items.

When sampling from a population, we cannot be sure that the distribution of values in the sample is normal. This is the case even if the population is normally distributed. However, there is a special situation where we are guaranteed a normal distribution. Specifically,

CENTRAL LIMIT THEOREM

For a given population, a distribution of means taken from random sets of independent samples of equal size are distributed normally.

To better understand the importance of the central limit theorem, let's consider the problem of determining the average American household credit card debt. There are approximately 127,000,000 American households. We have neither the time nor resources to poll each and every household. Therefore, we sample a random subset of 10,000 homes to obtain an average household credit card debt figure. We generalize our findings by reporting the obtained value as the average amount of American household credit card debt. An obvious question is: How confident can we be that the average computed from the sample data is an accurate estimate of the average household credit card debt for the general population?

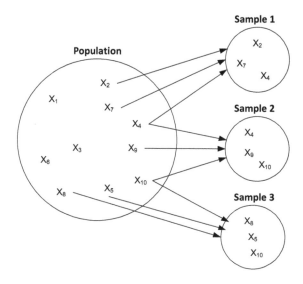

FIGURE 9.3   Random samples from a population of ten elements.

To help answer this question, suppose we repeat this experiment several times, each time recording the average household credit card debt for a new random sample of 10,000 homes. The central limit theorem tells us that the average values we obtain from the repeated process are normally distributed. Stated in a formal manner, we say that any one of the obtained sample means is an unbiased estimate of the mean for the general population. Also, the average of sample means taken over all possible samples of equal size is exactly equal to the population mean!

We now know that the average credit card debt computed from the initial sample mean of 10,000 households is an unbiased estimate of the average for the population. We still do not have a confidence in the computed average value. Although we cannot unequivocally state our computed value as an exact average debt figure for the general population, we can use the sample variance to easily obtain a confidence interval for the computed average value.

First, we estimate the variance of the population. The population variance is estimated by v/n, where v is the sample variance and n is the number of sample instances. Next we compute the *standard error*. The SE is simply the square root of the estimated population variance. The formula for computing SE is given as:

$$SE = \sqrt[2]{v/n} \tag{9.1}$$

As the population of sample means is normally distributed and the SE is an estimate of the population variance, we can make the following claim:

Any sample mean will vary less than plus or minus two SEs from the population mean 95% of the time.

For the household credit card debt problem, this means that we can be 95% confident that the actual average household credit card debt lies somewhere between two SEs above and two SEs below the computed sample mean. For our example, suppose the average household debt for our sample of 10,000 households is $6000 with an SE of 100. Our statement tells us that we can be 95% certain that the actual average American household credit card debt for the general population lies somewhere between $5800 and $6200.

We can use this technique as well as the hypothesis testing model introduced in the next section to help us compute confidence intervals for test set error rates, compare the classification error rates of two or more data mining models, and determine those numerical attributes best able to differentiate individual classes or clusters.

9.2.4 A Classical Model for Hypothesis Testing

You will recall that a hypothesis is an educated guess about the outcome of some event. Hypothesis testing is commonplace in the fields of science and medicine as well as in everyday situations dealing with politics and government. Let's take a look at a standard experimental design using a hypothetical example from the field of medicine.

Suppose we wish to determine the effects of a new drug, treatment X, that was developed to treat the symptoms associated with an allergy to house dust. For the experiment, we randomly choose two groups from a select population of individuals who suffer from the allergy. To perform the experiment, we pick one group as the *experimental group*. The second group is designated as the *control group*. Treatment X is applied to the experimental group and a placebo in the form of a sugar pill is distributed to the control group. We take care to make sure individual patients are not aware of their group membership. The average increase or decrease in the total number of allergic reactions per day is recorded for each patient in both groups. After a period of time, we apply a statistical test to determine if a significant difference in the measured parameter is seen between the two groups of patients.

A usual procedure is to state a hypothesis to be tested about the outcome of the experiment. Typically, the outcome is stated in the form of a *null hypothesis*. The null hypothesis takes a negative point of view in that it asserts any relationship found as a result of the experiment is due purely by chance. For our example, the null hypothesis declares the outcome will show no significant difference between the two groups of patients. A plausible null hypothesis for our experiment is

> *There is no significant difference in the mean increase or decrease of total allergic reactions per day between patients in the group receiving treatment X and patients in the group receiving the placebo.*

Notice that the null hypothesis specifies no significant difference rather than no significant improvement. The reason for this is that we have no *a priori* guarantee that treatment X will not cause an adverse reaction and worsen the symptoms associated with the allergy.

Once the experiment is performed and the data showing the results of the experiment have been gathered, we test if the outcome shows a significant difference between the two groups of patients. A classical model to test for a significant difference between the mean scores of a measured parameter such as the one for our experiment is one form of the well-known t-test given as

$$T = \frac{\left| \overline{x_1} - \overline{x_2} \right|}{\sqrt{v_1/n_1 + v_2/n_2}} \qquad (9.2)$$

where

T is the test statistic and
$\overline{x_1}$ and $\overline{x_2}$ are sample means for the independent t samples
v_1 and v_2 are variance scores for the respective means
n_1 and n_2 are corresponding sample sizes.

The term in the numerator is absolute difference between the two sample means. To test for a significant difference between the mean scores, the difference is divided by the SE for the distribution of mean differences. As the two samples are independent, the SE is simply the square root of the sum of the two variance scores associated with the two sample means.

TABLE 9.1 A Confusion Matrix for the Null Hypothesis

	Computed Accept	Computed Reject
Accept null hypothesis	True accept	Type 1 error
Reject null hypothesis	Type 2 error	True reject

To be 95% confident that the difference between two means is not due to chance, the value for T in the equation must be greater than or equal to 2 (see Figure 9.2). The model is valid because, like the distribution of means taken from sets of independent samples of equal size, the distribution of differences between sample means is also normal.

A 95% level of confidence still leaves room for error. With hypothesis testing two general types of error are defined. A *type 1 error* is seen when a true null hypothesis is rejected. A *type 2 error* is observed when a null hypothesis that should have been rejected is accepted. These two possibilities are shown in the confusion matrix of Table 9.1. For our experiment, a type 1 error would have us believing that treatment X has a significant impact on the average number of allergic reactions to dust when it does not. The degree of risk we are willing to take in committing a type 1 error is known as *statistical significance*. A type 2 error would state that treatment X does not affect the mean number of allergic reactions when indeed it does.

A requirement of the t-test described above is that each mean is computed from an independent dataset. Fortunately, many forms for the t-test exist. With machine learning, the usual case is to compare models with the help of a single test set of data. In Section 9.4, you will see how we employ a slight modification of the t-test given in Equation 9.2 to situations where we are limited to one test set.

9.3 COMPUTING TEST SET CONFIDENCE INTERVALS

Training and test data may be supplemented by *validation data*. One purpose of validation data is to help us choose one of several models built from the same training set. Once trained, each model is presented with the validation data, whereby the model showing a best classification correctness is chosen as the final model. Validation data can also be used to optimize the parameter settings of a supervised model so as to maximize classification correctness.

Once a supervised learner model has been validated, the model is evaluated by applying it to the test data. The most general measure of model performance is *classifier error rate*. Specifically,

$$Classifier\ Error\ Rate\ (E) = \frac{number\ of\ test\ set\ errors}{number\ of\ test\ set\ instances} \tag{9.3}$$

The purpose of a test set error rate computation is to give an indication as to the likely future performance of the model. How confident can we be that this error rate is a valid measure of actual model performance? To answer this question, we can use the SE statistic to calculate an error rate confidence interval for model performance. To apply the SE measure, we treat classifier error rate as a sample mean. Although the error rate is actually a proportion, if the number of test set instances is sufficiently large (say $n > 100$), the error rate can represent a mean value.

To determine a confidence interval for a computed error rate, we first calculate the SE associated with the classifier error rate. The SE together with the error rate is then used to compute the confidence interval. The procedure is as follows:

1. Given a test set sample S of size n and error rate E
2. Compute the sample variance as:

   $$\text{Variance}(E) = E(1 - E)$$

3. Compute the SE as the square root of Variance (E) divided by n.
4. Calculate an upper bound for the 95% confidence interval as $E + 2(SE)$.
5. Calculate a lower bound for the 95% confidence interval as $E - 2(SE)$.

Let's look at an example to better understand how the confidence interval is computed. Suppose a classifier shows a 10% error rate when applied to a random sample of 100 test set instances. We set $E = 0.10$ and compute the sample variance as

$$\text{Variance}(0.10) = 0.10(1 - 0.10) = 0.09$$

With a test set of 100 instances, the SE computation is

$$SE = \sqrt{(0.09/100)} = 0.03$$

We can be 95% confident that the actual test set error rate lies somewhere between two SEs below 0.10 and two SEs above 0.10. This tells us that the actual test set error rate falls between 0.04 and 0.16, which gives a test set accuracy between 84% and 96%.

If we increase the number of test set instances, we are able to decrease the size of the confidence range. Suppose we increase the test set size to 1000 instances. The SE becomes

$$SE = \sqrt{(0.09/1000)} \approx 0.0095$$

Making the same computations as in the previous example, the test set accuracy range is now between 88% and 92%. As you can see, the size of the test dataset has a marked effect on the range of the confidence interval. This is to be expected, because as the size of the test dataset becomes infinitely large, the SE measure approaches zero. Script 9.1 defines an R function that, given the test set error (or accuracy) rate and test set size, returns the upper and lower bounds for the 95% confidence interval.

Three general comments about this technique are as follows:

1. The confidence interval is valid only if the test data has been randomly chosen from the pool of all possible test set instances.

2. Test, training, and validation data must represent disjoint sets.

3. If possible, the instances in each class should be distributed in the training, validation, and test data as they are seen in the entire dataset.

Script 9.1 Computing Test Set Confidence Intervals

```
> model.conf <- function(error,n)
+ {
+    # Error is the test set error rate or accuracy.
+    # N represents the number of test set instances.
+    # This function returns the 95% confidence interval.
+    mvar <- error*(1-error)
+    SE = sqrt(mvar/ n)
+
+    # 95% confidence interval
+    up <- error + 2*SE
+    low <- error - 2*SE
+    return (c( up, low))
+    } # End Function

     # The function call. The error rate is 10% and n=100.
> model.conf(.10,100)

[1] 0.16 0.04
```

Chapter 1 showed you that test set error rate is but one of the several considerations when determining the value of a supervised learner model. To emphasize this point, let's assume that an average of 0.5% of all credit card purchases are fraudulent. A model designed to detect credit card fraud that always states a credit card purchase is valid will show a 99.5% accuracy. However, such a model is worthless as it is unable to perform the task for which it was designed. In contrast, a model that correctly detects all cases of credit card fraud at the expense of showing a high rate of a false positive classifications is of much more value.

One way to deal with this issue is to assign weights to incorrect classifications. With the credit card example, we could assign a large weight to incorrect classifications that allow a fraudulent card to go undetected and a smaller weight to the error of incorrectly identifying a credit card as fraudulent. In this way, a model will have its classification error rate increase if it shows a bias toward allowing fraudulent card usage to go undetected. In the next section, we show you how test set classification error rate can be employed to compare the performance of two supervised learner models.

9.4 COMPARING SUPERVISED MODELS

We can compare two supervised learner models constructed with the same training data by applying the classical hypothesis testing paradigm. Our measure of model performance is once again test set error rate. Let's state the problem in the form of a null hypothesis. Specifically,

There is no significant difference in the test set error rate of two supervised learner models, M_1 and M_2, built with the same training data.

Three possible test set scenarios are

1. The accuracy of the models is compared using two independent test sets randomly selected from a pool of sample data.

2. The same test set data is employed to compare the models. The comparison is based on a pairwise, instance-by-instance computation.

3. The same test data is used to compare the overall classification correctness of the models.

From a statistical perspective, the most straightforward approach is the first one, as we can directly apply the t statistic described in Section 9.2. This approach is feasible only if an ample supply of test set data are available. With large datasets, the prospect of extracting independent test set data is real. However, with smaller-sized data, a single test set may be the only possibility.

When the same test set is applied, one option is to perform an instance-by-instance pairwise matching of the test set results. This approach is described in Appendix B. Here we describe a simpler technique that compares the overall classification correctness of two models. The method can be applied to both the two independent (scenario 1) or single test set (scenario 3) cases. The most general form of the statistic for comparing the performance of two classifier models M_1 and M_2 is

$$T = \frac{\left| E_1 - E_2 \right|}{\sqrt{q(1-q)(1/n_1 + 1/n_2)}} \tag{9.4}$$

where
E_1 = The error rate for model M_1
E_2 = The error rate for model M_2
$q = (E_1 + E_2)/2$
n_1 = The number of instances in test set A
n_2 = The number of instances in test set B

Notice that $q(1-q)$ is a variance score computed using the average of the two error rates. With a single test set of size n, the formula simplifies to

$$T = \frac{\left| E_1 - E_2 \right|}{\sqrt{q(1-q)(2/n)}} \tag{9.5}$$

Script 9.2 defines a function that computes the value of T with a single test set. The script can be easily generalized for the two test set case. With either Equation 9.4 or 9.5, if the value of $T \geq 2$, we can be 95% confident that the difference in the test set performance of M_1 and M_2 is significant.

Script 9.2 Comparing Two Models Using the Same Test Data

```
> model.Compare <- function (E1, E2, n)
+ {
+    # E1 and E2 are error rates for the two models
+    # n is the number of test set instances
+    # Compute the variance
+    q <- (E1 + E2)/2
+    var <- q* (1 - q)
+
+    # Compute the t value
+    den <- 1 / n
+    P<- abs(E1 - E2)/sqrt(var*2*den)
+
+    return(round(P,3))}

# The function call n=100, E1=20%, E2=30%
> model.Compare (.2,.3,100)

[1] 1.633
```

9.4.1 Comparing the Performance of Two Models

Let's look at an example. Suppose we wish to compare the test set performance of learner models M_1 and M_2. We test M_1 on test set A and M_2 on test set B. Each test set contains 100 instances. M_1 achieves an 80% classification accuracy with set A, and M_2 obtains a 70% accuracy with test set B. We wish to know if model M_1 has performed significantly better than model M_2. The computations are

- For model M_1: $E_1 = 0.20$

- For model M_2: $E_2 = 0.30$

- q is computed as
  $(0.20 + 0.30)/2 = 0.25$

- The combined variance $q(1 - q)$ is
  $0.25(1.0 - 0.25) = 0.1875$

- The computation for P is

$$T = \frac{|0.20 - 0.30|}{\sqrt{0.1875(1/100 + 1/100)}}$$

$$T \approx 1.633$$

As $T < 2$, the difference in model performance is not considered to be significant. We can increase our confidence in the result by switching the two test sets and repeating the experiment. This is especially important if a significant difference is seen with the initial test set selection. The average of the two values for T is then used for the significance test.

9.4.2 Comparing the Performance of Two or More Models

The various forms of the t-test are limited to comparing pairs of competing models. Using multiple t-tests is one way around this problem. For example, six t-tests can be used to compare four competing models. Unfortunately, in the world of statistics, it is a well-known fact that performing multiple t-tests increases the chance of a type 1 error. Because of this, a one-way analysis of variance (ANOVA) is often used in situations when two or more model comparisons are needed.

A one-way ANOVA uses a single factor to compare the effects of one treatment/independent variable on a continuous dependent variable. For example, suppose we wish to test for a significant difference in the number of hours of TV watched by individuals in three age groups:

- *age <= 20*

- *age > 20* and *age <=40*

- *age > 40*

For this example, the dependent variable is the number of hours of TV watched per week and the independent variable is *age group.*

The ANOVA uses the F statistic—introduced in Chapter 5—significant differences between the groups defined by the independent variable. The F statistic is computed as the ratio of between group variance divided by within group variance. For our example, the computed F ratio represents variance in the number of hours of TV watched between the three age groups relative to the variance in the number of hours watched within each of the three groups.

The F ratio increases as between group variance increases when compared to within group variance. The larger the F ratio, the greater the likelihood of a significant difference between the means of the tested groups. A table of critical values determines if an F ratio is statistically significant.

When the ANOVA is applied to test for significant differences in the performance of two or several data mining models, the model becomes the independent variable. Each tested model is one instance of the independent variable. The dependent variable is model error rate.

Although the ANOVA is the stronger test, it is important to note that when three or more models are tested, the F statistic tells whether any of the models differ significantly in performance but does not indicate exactly where these differences lie.

Lastly, the t-test and the ANOVA are known as parametric statistical tests as they make certain assumptions about the parameters (mean and variance) they use to describe the characteristics of the data. Two such assumptions are that the choice of one sample does not affect the chances of any other sample being included and that the data is taken from normally distributed populations having equal variances. However, the tests are often used even when these assumptions cannot be completely verified.

9.5 CONFIDENCE INTERVALS FOR NUMERIC OUTPUT

Just as when the output was categorical, we are interested in computing confidence intervals for one or more numeric measures. For purposes of illustration, we will use *mean absolute error*. As with classifier error rate, *mean absolute error* is treated as a sample mean. The sample variance is given by the formula:

$$variance(mae) = \frac{1}{n-1}\sum_{i=1}^{n}(e_i - mae)^2 \tag{9.6}$$

where
e_i is the absolute error for the ith instance
n is the number of instances.

Next, as with classifier error rate, we compute the SE for the *mae* as the square root of the variance divided by the number of sample instances.

$$SE = \sqrt{\frac{variance(mae)}{n}} \tag{9.7}$$

Finally, we calculate the 95% confidence interval by respectively subtracting and adding two SEs to the computed *mae*. That is,

Upper error bound = $mae + 2SE$

Lower error bound = $mae - 2SE$

The *mmr.stats* function can be easily modified to include confidence intervals for *mae*—see R programming Exercise 1. Finally, if your interests in numeric model testing lie beyond our discussion here, equations and examples for comparing supervised learner models having numeric output can be found in Appendix B.

9.6 CHAPTER SUMMARY

A model's performance is influenced by several components, including training data, input attributes, learner technique, and learner parameter settings to name just a few. Each component that in some way dictates model performance is a candidate for evaluation.

Supervised model performance is most often evaluated with some measure of test set error. For models having categorical output, the measure is test set error rate computed as the ratio of test set errors to total test set instances. For models whose output is numeric, the error measure is usually the mean squared error, the root mean squared error, or the mean absolute error.

Although most data is not normally distributed, the distribution of sample means taken from a set of independent samples of the same size is distributed normally. We can take advantage of this fact by treating test set error rate as a sample mean and applying the properties of normal distributions to compute test set error confidence intervals. We can also apply classical hypothesis testing to compare test set error values for two or more supervised learner models. These statistical techniques offer us a means to associate measures of confidence with the output of our data mining sessions.

9.7 KEY TERMS

- *Classifier error rate.* The number of test set errors divided by the number of test set instances.

- *Control group.* In an experiment, the group not receiving the treatment being measured. The control group is used as a benchmark for measuring change in the group receiving the experimental treatment.

- *Experimental group.* In a controlled experiment, the group receiving the treatment whose effect is being measured.

- *Mean.* The average of a set of numerical values.

- *Mean absolute error.* Given a set of instances each with a specified numeric output, the mean absolute error is the average of the sum of absolute differences between the classifier predicted output for each instance and the actual output.

- *Mean squared error.* Given a set of instances each with a specified numeric output, the mean square error is the average of the sum of squared differences between the classifier predicted output and actual output.

- *Normal distribution.* A distribution of data is considered normal if a frequency graph of the data shows a bell-shaped or symmetrical characteristic.

- *Null hypothesis.* The hypothesis of no significant difference.

- *Root mean squared error.* The square root of the mean squared error.

- *Sample data.* Individual data items drawn from a population of instances.

- *Standard deviation.* The square root of the variance.

- *Standard error.* The square root of a sample variance divided by the total number of sample instances.

- *Stratification*. Selecting data in a way so as to ensure that each class is properly represented in both the training and test set.

- *Statistical significance*. The degree of risk we are willing to take in committing a type 1 error.

- *Type 1 error*. Rejecting a null hypothesis when it is true.

- *Type 2 error*. Accepting a null hypothesis when it is false.

- *Validation data*. A set of data that is applied to optimize parameter settings for a supervised model or to help choose from one of several models built with the same training data.

- *Variance*. The average squared deviation from the mean.

EXERCISES

Review Questions

1. Differentiate between the following terms:

   a. Validation data and test set data

   b. Type 1 and type 2 errors

   c. Control group and experimental group

   d. Mean squared error and mean absolute error

2. For each of the following scenarios, state the type 1 and type 2 errors. Also, decide whether a model that commits fewer type 1 or type 2 errors would be a best choice. Justify each answer.

   a. A model for predicting if it will snow

   b. A model for selecting customers likely to purchase a television

   c. A model to decide likely telemarketing candidates

   d. A model to predict whether a person should have back surgery

   e. A model to determine if a tax return is fraudulent

3. When a population is normally distributed, we can be more than 95% confident that any value chosen from the population will be within two standard deviations of the population mean. What is the confidence level that any value will fall within three standard deviations on either side of the population mean?

Programming with R

1. Modify mmr.stats to include computing and printing confidence intervals for mae. Here's some help:

   a. Initialize a variable (varTotal) at 0 whose purpose is to keep a running total of the sum of squared absolute errors.

   b. Make abs.error a subscripted variable as individual error values must be saved to compute variance.

   c. After mae has been computed, create a loop to accumulate the squared errors. That is,

   varTotal = varTotal + (abs.error[i] - mae)*(abs.error[i]-mae)

   d. Compute the variance by multiplying varTotal by 1/(n-1)

   e. Use varTotal to compute SE and the error bounds. Print the lower and upper limits for the error.

   f. Test your modification to mmr.stats by executing Script 5.3.

2. Modify mmr.stats to include confidence limits for the root mean squared error and the residual SE. Test your modification to mmr.stats by executing Script 5.3.

Installed Packages and Functions

Package Name	Function(s)
base / stats	abs, c, function, return, round, sqrt

## CHAPTER 10

Support Vector Machines

In This Chapter

- Creating Support Vector Machines (SVMs)

- Evaluating Support Vector Models

- Applying SVMs to Microarray Data

$\mathbf{S}$UPPORT VECTOR MACHINES (SVMs) provide a unique approach for classifying both linearly separable and nonlinear data. The first SVM algorithm was developed by Vladimir Vapnik and Alexey Chervonenkis in the 1960s. However, SVMs drew general interest only after their formal introduction to the research community in 1992 (Boser et al., 1992). SVMs have been applied in several areas including time series analysis, bioinformatics, textual data mining, and speech recognition. Unlike backpropagation neural networks and other techniques that often create locally optimal models, SVMs are able to provide globally optimal solutions. In addition, because SVMs use the most difficult to classify instances as the basis for their predictions, they are less likely to over fit the data.

The basic idea of how an SVM algorithm works is conceptually simple, but the mathematics can be a challenge. To help remove much of the mystery behind SVMs, we take an example-based approach.

Although SVMs can be used for both classification and numeric prediction, our focus is on SVMs that create binary classification models. Those who desire to delve deeper into the more theoretical aspects of SVMs can find a wealth of information both in textbooks (Vapnik, 1998, 1999) and on the Web.

The key to the workings of an SVM is the hyperplane. A hyperplane is simply a subspace one dimension less than its ambient space. To see this, consider Figure 10.1 which displays a two-dimensional space with linearly separable classes. One class is represented by stars and the second by circles. In two dimensions, a hyperplane is a straight line. The figure shows two such hyperplanes labeled h_1 and h_2. Clearly, there is an infinite number of hyperplanes dividing the classes. So, which hyperplane is the best choice? It turns out that

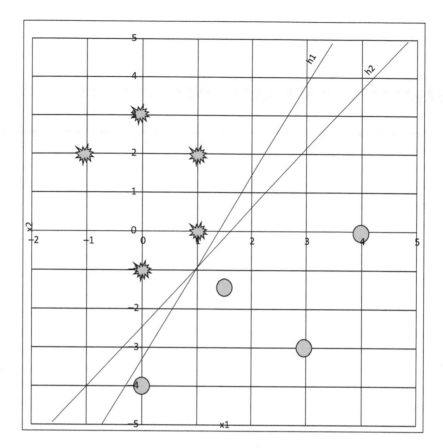

FIGURE 10.1 Hyperplanes separating the circle and star classes.

it's the hyperplane showing the greatest separation between the class instances. This particular hyperplane is unique and has a special name. It is known as the *maximum margin hyperplane* (MMH). Given a dataset having two classes, the job of the SVM algorithm is to find the MMH. Here's how it's done.

An SVM uses the instances lying on the outer marginal boundaries between the classes to find the MMH. To see this, consider Figure 10.2 which shows two hyperplanes each passing through the instances defining the outer margin boundaries for their respective classes. The hyperplane labeled h_1 passes through the two boundary instances for the star class and h_2 passes through the single boundary instance for the circle class. These boundary instances are known as *support vectors*. The support vectors associated with the MMH have the best chance of differentiating between the classes as they lie along the largest marginal instance gap. The SVM algorithm builds its model using only these critical support vectors. All other class instances are irrelevant. Here is a general description of the process used by the SVM algorithm:

- Determine if the instances defining the classes are linearly separable in the current dimension. Provided that the classes are separated by a linear boundary, the support vectors are used to establish the MMH.

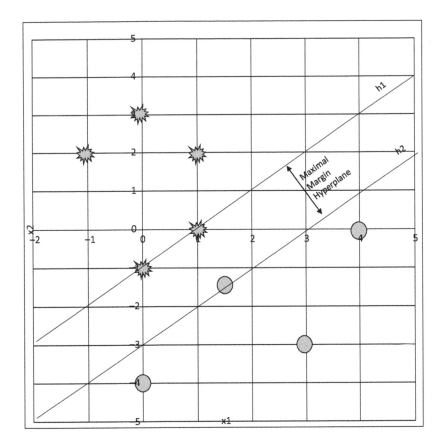

FIGURE 10.2 Hyperplanes passing through their respective support vectors.

- If the classes are not linearly separable, map the instances into consecutively higher dimensions until linear separability is achieved. Once again, the support vectors found in the new higher dimension determine the MMH.

Finding the support vectors and MMH is a problem in constrained quadratic optimization, the details of which are beyond the scope of this book. The important point is that given certain constraints, the MMH and support vectors can be determined in a reasonable amount of time.

In the next section, we present a simple two-dimensional example with linearly separable data. We follow our example with a discussion of the nonlinear case where we detail the dimensional mapping process. We complete our study of SVMs with several experiments using the SVM software available with R.

10.1 LINEARLY SEPARABLE CLASSES

When discussing SVMs, it is useful to reference instances as vectors. A *vector* is a quantity having both direction and magnitude. Instances can be displayed horizontally as row vectors or vertically as a column vectors. We denote vectors in bold face type.

One vector operation of particular importance is the *dot product*. The dot product can be thought of as a measure of similarity. Algebraically, given two vectors of equal length, the dot product of two vectors is the sum of the products of their corresponding entries. For example, given $\mathbf{v} = (v_1, v_2)$ and $\mathbf{w} = (w_1, w_2)$, then $\mathbf{v} \cdot \mathbf{w} = (v_1 w_1 + v_2 w_2)$. It is common to see the previous equation written as $\mathbf{v} \cdot \mathbf{w}^T = (v_1 w_1 + v_2 w_2)$ where $\mathbf{v}$ is the row vector and $\mathbf{w}^T$ is the transpose of row vector $\mathbf{w}$. This information together with Figure 10.2 gives us what we need for our first example!

The coordinate axes in Figure 10.2 are labeled x_1 and x_2 rather than x and y. The rationale is that x_1 and x_2 are both independent, whereas the class associated with each vector is dependent. Here each vector instance is a member of one of two classes. The class marked by stars contains instances (0, –1), (0, 3), (1, 0), (1, 2), and (–1, 2). The circle class shows instances (0, –4), (1.5, –1.5), (3, –3), and (4, 0). We have already visually determined the classes to be linearly separable. Our goal is to find the MMH separating the classes. To find the MMH, we must locate the support vectors. For our simple example, this task is easy.

Figure 10.2 shows the two support vectors for the star class to be (1, 0) and (0, –1). The figure also shows (1.5, –1.5) as the lone support vector for the circle class. If there is any doubt about which instances represent support vectors, we can use simple Euclidean distance to either determine or verify our choices. For this example, the Euclidean distance between (1.5, –1.5) and (1, 0) is $\sqrt{2.5}$ as is the distance between (1.5, –1.5) and (0, –1). As all other between-class vector distances are greater than $\sqrt{2.5}$, only these three vectors are relevant to the SVM algorithm.

The class with support vectors (1, 0) and (0, –1) clearly lies above the margin of separation. By convention, this class is denoted as the positive class and is given a label of 1. The support vector (1.5, –1.5) represents the remaining class and lies below the class margin. Therefore, (1.5, –1.5) is a negative example. Convention tells us to give this class a label of –1. Although this labeling scheme is standard, choosing the values to represent each respective class is arbitrary other than having all support vectors on upper part of the margin get the same positive value and all vectors on the lower margin receive the corresponding negative value.

Next, establishing the MMH requires a general form for a two-dimensional equation. We have two choices. Here is a standard representation of a linear equation in two dimensions with a constant (bias) term included.

$$w_0 + w_1 x_1 + w_2 x_2 = 0 \tag{10.1}$$

where
x_1 and x_2 are attribute values
w_0, w_1, and w_2 are parameters to be learned.

However, if the support vectors are modified by adding a term to account for the bias, we can also write the MMH in terms of its support vectors in the following way:

$$\sum a_i x_i \cdot x = 0 \tag{10.2}$$

where
    a_i is the learned parameter associated with the ith support vector
    x_i is the ith support vector
    x is the vector instance to be classified.

Equation 10.2 better suits our needs for determining the MMH. To use Equation 10.2, we first supplement each support vector with a bias of 1 to account for the constant term in the equation. To match Equation 10.1, we add the bias as the first term. The modified class 1 support vectors now appear as $SV_1 = (1, 1, 0)$ and $SV_2 = (1, 0, -1)$. The modified support vector for the class labeled -1 is $SV_3 = (1, 1.5, -1.5)$. Knowing that x in Equation 10.2 represents an instance to be classified, the assignment of $a_0 = w_0$, $a_1 = w_1$, and $a_2 = w_2$ confirms the equivalency of Equations 10.1 and 10.2.

We need one more equation. Specifically, a vector representation for the parameters making up the MMH—w_0 w_1 and w_2. The parameters can be given in vector notation as

$$w = \sum a_i sv_i \tag{10.3}$$

where the
    a_i values are to be learned
    sv_i represents the ith support vector.

With three support vectors and the above equations, determining the MMH is a matter of solving three equations in three unknowns.

Consider the augmented support vector $x = (1, 1, 0)$. Using 10.2 and the fact that $(1, 1, 0)$ lies on the hyperplane represented by positive 1, it must satisfy this condition.

$$a_1 (1, 1, 0) \cdot (1, 1, 0)^T + a_2 (1, 0, -1) \cdot (1, 1, 0)^T + a_3 (1, 1.5, -1.5) \cdot (1, 1, 0)^T = 1$$

In a like manner, $(1, 0, -1)$ must satisfy

$$a_1 (1, 1, 0) \cdot (1, 0, -1)^T + a_2 (1, 0, -1) \cdot (1, 0, -1)^T + a_3 (1, 1.5, -1.5) \cdot (1, 0, -1)^T = 1$$

Finally, $(1, 1.5, -1.5)$ is on the hyperplane of negative examples so it must satisfy

$$a_1 (1, 1, 0) \cdot (1, 1.5, -1.5)^T + a_2 (1, 0, -1) \cdot (1, 1.5, -1.5)^T + a_3 (1, 1.5, -1.5) \cdot (1, 1.5, -1.5)^T = -1$$

Computing the dot products, we have

$$2a_1 + a_2 + 2.5a_3 = 1$$

$$a_1 + 2a_2 + 2.5a_3 = 1$$

$$2.5a_1 + 2.5a_2 + 5.5a_3 = -1$$

Solving the system of equations gives us

$$a_1 = 2 \quad a_2 = 2 \quad a_3 = -2$$

With values for a_1, a_2, and a_3, we use Equation 10.8 to determine w_0, w_1, and w_2.

$$(w_0, w_1, w_2) = 2(1,1,0) + 2(1,0,-1) - 2(1,1.5,-1.5)$$

This gives us

$$w_0 = 2, w_1 = -1 \text{ and } w_2 = 1.$$

Therefore, the equation for the MMH is

$$2 - x_1 + x_2 = 0 \tag{10.4}$$

Figure 10.3 displays the equation of the MMH together with the equations for the two boundary hyperplanes. Using the general form of the equations for the decision boundaries, it can be shown that the distance d between the two hyperplanes is given by $2/\|\mathbf{w}\|$

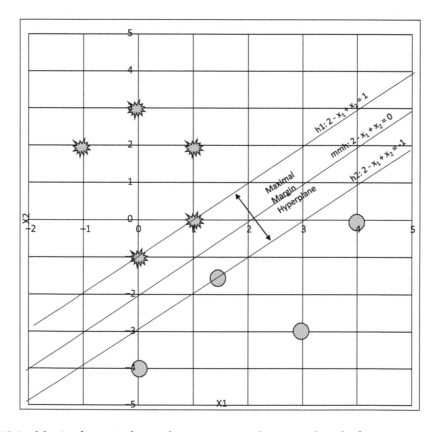

FIGURE 10.3   Maximal margin hyperplane separating the star and circle classes.

where $\|\mathbf{w}\|$ is the Euclidean norm of $\mathbf{w}$ defined as the square root of the dot product $\mathbf{w}\cdot\mathbf{w}$. That is,

$$\|w\| = \sqrt{\left(W_1^2 + W_2^2 + \cdots + W_n^2\right)} \tag{10.5}$$

During training, the support vector algorithm begins by estimating the parameters of the two decision boundaries under certain constraints. One obvious constraint is the maximization of this margin. For our example, Equation 10.5 tells us that the width of the maximal margin = $\sqrt{(w_1^2 + w_2^2)}$ which computes to $\sqrt{2}$.

Knowing the support vectors and the MMH, we can apply our model to new instances for classification. Two alternatives present themselves. We can use the MMH by placing the x_1 and x_2 values of an unclassified instance into Equation 10.4. If the result is greater than 0, we classify the instance with class label 1. If the value is negative, it is placed with the class labeled –1. Alternatively, we can use the equation for the MMH in terms of the support vectors, but this time, with the bias included as the constant term. Specifically,

$$y = b + \sum a_i x_i \cdot x \tag{10.6}$$

where
   a_i is the learned parameter associated with the ith support vector
   b is the learned constant term
   x_i is the ith support vector
   x is the vector instance to be classified.

If y is positive, we classify an unknown instance with class label 1. Likewise, if y proves to be negative, the instance is associated with the class labeled –1. If $y = 0$, the instance lies on the MMH and is classified with the most commonly occurring class. Let's use Equation 10.6 to determine the classification of the instance (0.5, 0).

To make our computation, we remove the bias term from our support vectors and include the value 2 for b. Applying the previously computed values for a_1, a_2, and a_3, we have

$$y = 2 + 2\left[(1,0)\cdot(0.5,0)^{\mathrm{T}}\right] + 2\left[(0,-1)\cdot(0.5,0)^{\mathrm{T}}\right] - 2\left[(1.5,-1.5)\cdot(0.5,0)^{\mathrm{T}}\right]$$

This gives us a value of 1.5 for y which tells us that (0.5, 0) lies above the MMH and is classified with the vectors labeled 1. Using Equation 10.6 to classify (0, –5) gives a value of –3 telling us (0, –5) is classified with the class labeled –1.

Using 10.6 to classify support vector (1, 0) gives a value of 1. This is expected as (1, 0) lies on hyperplane h_1. Likewise applying Equation 10.6 to (1.5, –1.5) offers a result of –1. Using the model to classify (3, 1) gives a value of 0. This tells us (3, 1) is on the MMH.

Before we proceed to the nonlinear case, it is obvious that a linear model is preferred especially if an acceptable level of accuracy is seen. That said, it is worth noting that if the classes are almost linearly separable in the current dimension, a slight modification of the SVM algorithm gives it the capability of creating a linear model.

10.2 THE NONLINEAR CASE

The strength of SVMs lies in their ability to classify in clearly nonlinear domains. When the instances in the current dimensional space are not linearly separable, an SVM applies a nonlinear mapping, thereby transforming the instances into a new, likely higher dimension. An unacceptable degree of complexity can enter the picture unless the transformation is accomplished with a mapping function having certain special characteristics. Let's see why this is the case.

Consider the two-dimensional vectors **x** and **z** with respective attribute values (x_1, x_2) and (z_1, z_2). Suppose that in an attempt to achieve linearity, our function transforms the current two-dimensional space by applying the mapping

$$(x_1, x_2) \rightarrow (x_1, x_2, x_1^2, x_1 x_2, x_2^2)$$

Prior to the above mapping, the dot product computation for each vector pair requires one multiplication and one addition. In the new space, the dot product of each vector pair **x** and **z** entails five multiplications and four additions. Specifically,

$$\mathbf{x} \cdot \mathbf{z} = (x_1 z_1, x_2 z_2, x_1^2 z_1^2, x_1 x_2 z_1 z_2, x_2^2 z_2^2)$$

Once a model has been established, the model's application also requires the same dot product computations. However, given the right type of mapping, these added computations can be completely avoided! The process requires using a special mapping function known as a *kernel function*. For a function K to qualify as a kernel function, the mapping function ϕ must be chosen such that

$$K(x, z) = \phi(x) \cdot \phi(z) \tag{10.7}$$

To illustrate a function with this property, consider the mapping ϕ with $\mathbf{x} = (x_1, x_2)$

$$\phi(x_1, x_2) = \left(1, \sqrt{2}x_1, \sqrt{2}x_2, x_1^2, x_2^2, \sqrt{2}x_1 x_2\right)$$

Computing the dot product of $\phi(x_1, x_2)$ and $\phi(z_1, z_2^T)$ and factoring gives us

$$\phi(x_1, x_2) \cdot \phi(z_1, z_2) = \left(1 + x_1 z_1 + x_2 z_2\right)^2$$

This tells us that if we assign

$$K(x, z) = \left(1 + x_1 z_1 + x_2 z_2\right)^2 \tag{10.8}$$

the property stated in Equation 10.7 is satisfied and K qualifies as a kernel function. More importantly, it tells us that the dot product computations can be performed in the original space rather than in the higher dimension! The general form of Equation 10.8 is known as the polynomial kernel of degree n and is given as

$$K(x, z) = \left(1 + x_1 z_1 + x_2 z_2\right)^n \qquad (10.9)$$

In general, every location within the training algorithm where the dot product representation $\phi(\mathbf{x}) \cdot \phi(\mathbf{z})$ appears can be replaced by $K(\mathbf{x}, \mathbf{z})$. This is true for both model training and application. In addition to the polynomial function, several other kernel functions have been studied some of which are built into the SVM models available with R.

10.3 EXPERIMENTING WITH LINEARLY SEPARABLE DATA

Our first experiment uses John Platt's (1998) sequential minimum optimization algorithm (SMO) contained in the RWeka package. SMO is a supervised learner that automatically transforms categorical attributes into binary values and by default omits any instances with missing values. It also normalizes all attributes by default. SMO invokes the polynomial kernel discussed above, but others can be used. The one-vs-one method is applied in domains with more than two classes.

The purpose of this experiment is to verify the correctness of the linearly separable example given above. That is, we will apply SMO to the data given in Figure 10.1 and compare the resultant equation with the one shown in Equation 10.4. If our work is correct, the two equations should match. The data points are contained in the file SVM10.1.csv. Here's what to do:

- Import SVM10.1.csv into the RStudio environment.

- Use str to make sure SVM10.1 is a data frame with output variable of type factor.

- Load the RWeka package (library RWeka).

- Type WOW(SMO) to list the options for SMO.

The options (not displayed) show that setting $N = 2$ disables data normalization. Data normalization is the rule. However, because the input attributes for our example are of the same scale, data normalization is not needed. In fact, unnecessary normalization with small datasets can be detrimental.

- Load and execute Script 10.1 to obtain the output shown in the script.

Script 10.1 SMO Applied to the Data Points in Figure 10.1

```
> library(RWeka)
> my.svm <- SMO(class ~ x1+x2, data =SVM10.1,control=
+       Weka_control(N=2))

> # Print out the equation of the MMH
> my.svm
SMO
```

```
Kernel used:
  Linear Kernel: K(x,y) = <x,y>

Classifier for classes: one, zero

BinarySMO

Machine linear: showing attribute weights, not support vectors.

          1        * x1
  +      -1        * x2
  -       2


Number of kernel evaluations: 21 (84.783% cached)


> # Output the confusion matrix
> summary(my.svm)

=== Summary ===

Correctly Classified Instances          9                100      %
Total Number of Instances               9

=== Confusion Matrix ===

 a b    <-- classified as
 5 0 |  a = one
 0 4 |  b = zero
```

The output displayed in Script 10.1 makes it clear that the linear model created by SMO is identical to Equation 10.4. The confusion matrix tells us that the MMH correctly classified each data point. You will learn much more about SMO while exploring the end-of-chapter exercises.

For our second experiment, the dataset remains the same, but this time we use *svm*—the SVM function housed in the *package e1071*—to verify our result. What's that old saying? Verify twice then take it to the boss! Be sure to install package *e1071* prior to typing the call to *svm* seen in Script 10.2.

As with SOM, the default is to normalize the data. Scale = FALSE inhibits this normalization. Also, the kernel specification is necessary as *svm's* default kernel is the radial basis function. The variable *na.action* (not shown) defaults to *na.omit*. That is, any instance with one or more NA values is deleted. As our very small dataset does not contain missing items, we need not concern ourselves with setting the value of *na.action*.

As you can see, the output displayed for Script 10.2 verifies Equation 10.4. In the final sections of this chapter, we examine a much more interesting problem that deals with nonlinear data!

Script 10.2 *svm* Applied to the Data Points in Figure 10.1

```
> library(e1071)
> my.svm <- svm(class~ .,data =SVM10.1,scale = FALSE,
kernel="linear")
> my.svm

Call:
svm(formula =class ~ ., data = SVM10.1,kernel ="linear", scale
=FALSE)


Parameters:
   SVM-Type:  C-classification
 SVM-Kernel:  linear
       cost:  1

Number of Support Vectors:   3

> coef(my.svm)
(Intercept)           x1              x2
          2           -1               1
```

10.4 MICROARRAY DATA MINING

Two women are diagnosed with stage 3 breast cancer, both receive the same treatment. One woman recovers to live another 20 years, the second woman passes in a matter of weeks. This is a common story for most types of cancer afflicting both men and women. One of the reasons for this is that the cancer cells of the two individuals with the same diagnosis may look the same under a microscope but at the molecular level, are very different. These molecular level differences can cause one incidence of cancer to metastasize and another to respond well to treatment.

Researchers in the field of bioinformatics—or computational biology—study these molecular level differences for many reasons including: helping doctors develop treatment plans for individual patients, early disease detection, new drug development, and prediction of treatment outcome to name a few. A detailed discussion of how this is done is well beyond the scope of this text. However, an overview of the very basic biology of this process will help you better understand what microarray data mining is all about.

10.4.1 DNA and Gene Expression

Humans have trillions of cells all originating from a single cell (the fertilized egg). Cells are of many types: Skin, nerve, blood, brain. Each cell contains a complete copy of the *program* for making the organism. This program is known as the genome. The genome is encoded in DNA and represents the set of all genes in a given organism.

A *gene* is a contiguous subpart of DNA. Human DNA contains about 30–35,000 genes. A gene is expressed—converted into a functional product—by manufacturing messenger RNA (mRNA) which is later translated into a protein. That is, gene expression acts as an on/off switch to control the manufacturing of protein. At any given point in time, approximately 40% of human genes are expressed.

Microarrays—also known as DNA chips—are able to measure this level of mRNA expression.

A *microarray* is actually a collection of microscopic DNA spots attached to a solid surface. The gene expression levels for a particular individual obtained from a microarray represent the molecular level differences that can be used to help determine individual patient outcome and treatment.

Microarray data mining is the process of analyzing the gene expression data obtained from microarray devices. Microarray data mining differs from traditional data mining in that a typical application will contain several thousand attributes (genes) but few instances. It is easy to see that a defining problem for microarray mining is attribute selection/elimination. Let's take a look at the attribute selection process.

10.4.2 Preprocessing Microarray Data: Attribute Selection

Microarray data preprocessing takes place in two stages. The first stage is independent of any classes within the data.

10.4.2.1 Class-Independent Preprocessing

Common first-stage techniques include thresholding, normalization and filtering. *Thresholding* involves setting an upper and lower bound value for gene attributes. Genes that do not fall within the specified boundary are not analyzed. A best *normalization* method is to normalize data to have a mean of zero and a standard deviation of one. *Filtering* methods remove genes that display little variation. For example, genes that have a range less than a fixed value are removed.

10.4.2.2 Class-Dependent Preprocessing

The purpose of stage-two preprocessing is to develop a best set of genes for each class within the data. As stage-two preprocessing is class dependent, it only applies to supervised learning.

10.4.2.2.1 The Wrapper Approach

The *wrapper* approach selects the top 100–200 genes per class ordered by their within class rank. Models are created using various numbers of genes per class. Each model is tested using a cross validation method. The best-performing model is chosen as the final representation of the data.

10.4.2.2.2 Permutation Testing

A small number of instances and a large number of attributes can show *false positive* correlations between input genes and the output classes. *Permutation testing*—sometimes called randomization—is used to help identify these false positive correlations. A main advantage of this technique is its ability to estimate the significance levels of several genes in parallel. A second advantage is seen in that it works well even with very small datasets. The method is most easily explained assuming two classes and a single gene. Specifically,

1. Compute the t score reflecting the between-class mean expression level difference for the gene.

2. Leave the instances in their current order but randomly permute the output attribute. That is, each gene instance is randomly assigned to one of the two classes.

3. Recompute and record the t score for the dataset created in 2.

4. Repeat steps 2 and 3 hundreds or thousands of times.

Once the total of all trials is complete, the *t* scores are arranged empirically in ascending order. If the original t score falls outside of the 95% range of all t scores, we can be 95% certain that the difference in gene expression level between the two classes is significant.

Permutation testing is a widely accepted approach for dealing with skewed distributions often seen with small-sized datasets. You can experiment with permutation testing by installing R's *coin* and *lmPerm* packages.

10.4.3 Microarray Data Mining: Issues

The main problem seen with most microarray applications is the amount of preprocessing necessary in order to obtain acceptable results. A second issue is that the imbalance between the number of attributes (genes) and the number of instances often leads to misclassifications. A third problem is the possibility of collecting noisy data. A fourth concern is realized when there is a lack of dominant genes able to determine outcomes. A fifth problem is that models are often unable to explain their results.

Lastly, most machine learning techniques do not perform well when the number of attributes greatly exceeds the number of instances. The good news is that SVMs are an exception to the rule! This is the case as SVMs do not require large amounts of training data to build accurate models as they focus solely on those instances that lie along class boarders. Also, SVMs are able to deal with the high dimensionality often seen with microarray applications.

10.5 A MICROARRAY APPLICATION

With this basic overview behind us, let's take a look at some data! The dataset we will use for our experiments—*SBRCTDNAData*—is stored in both .csv and MS Excel format in your data directory. The data contains 59 attributes representing gene expression levels for 88 cancer patients. Each instance characterizes a patient afflicted with small round blue cell tumors (SBRCT) observed during childhood. The original dataset (Khan et al., 2001) consists of 2308 attributes. The researchers used their attribute reduction method to identify 59 of the most significant input attributes. The tumor types defined in the data include

Burkitt lymphoma (BL), Ewing sarcoma (EWS), neuroblastoma (NB), and rhabdomyosarcoma (RMS). These cancer types are difficult to distinguish under the microscope, and no single test can precisely differentiate them. Accurate diagnosis of SRBCTs is essential because prognoses and treatment options vary widely with each type. Please note that five of the instances within the training data are not part of the four classes mentioned above. The *description* sheet within the MS Excel version of the dataset provides additional information about the data and its use.

Our goal is to determine how well a model trained with *svm* is able to determine test data tumor type. Additionally, we want to see what happens to test set accuracy when subsets of the 59 attributes are used for the model building process.

10.5.1 Establishing a Benchmark

Script 10.3 shows the process and partial output of our first experiment. During data preparation, the first column within the dataset is eliminated as it holds a unique identifier for each row. The training data contains the first 61 instances with the remaining instances reserved for testing. The call to sbr.perform produces an 8×8 confusion matrix (not displayed).

Script 10.3 Applying *svm* to the SBRCT Cancer Patient Dataset

```
> library(e1071)

> # Prepare the data
> sbr.data <- SBRCTDNAData[,-1]
> sbr.train <- sbr.data[1:61,]
> sbr.test <- sbr.data[62:88,]

> # Build the model
> sbr.model <-svm(Class ~ ., data=sbr.train)
> summary(sbr.model)

Call:
svm(formula = Class ~ ., data = sbr.train)

Parameters:
   SVM-Type:  C-classification
 SVM-Kernel:  radial
       cost:  1

Number of Support Vectors:  51

Number of Classes:  8

Levels:
```

```
  BL EWS NB Osteosarcoma Prostate Ca. RMS Sarcoma Sk. Muscle

> # Test the model
> sbr.pred <- predict(sbr.model,sbr.test)
> sbr.perform <- table(sbr.test$Class,
+                      sbr.pred,dnn=c("Actual", "Predicted"))
> sbr.perform

# The 8X8 confusion matrix will display here.

> confusionP(sbr.perform) #Computes accuracy.

  Correct= 21 Incorrect= 6
  Accuracy = 77.78 %
```

The result of our experiment shows that 21 of 27 (77.78%) test set instances were correctly identified. If you examine the confusion matrix, you will see the incorrect classifications include four NB instances classified as EWS instances, and two RMS instances given an EWS label. With this benchmark established, let's use a simple attribute elimination technique to see if we can do better with fewer input attributes.

10.5.2 Attribute Elimination

As these data have been preprocessed, our chance of producing a better model having fewer attributes is questionable. However, we can't let a negative attitude turn us away from at least giving it a try! Let's perform a simple attribute elimination by first ranking the attributes from most to least useful. We then proceed to build a benchmark model using all of the attributes. The second model is constructed after eliminating the least-effective individual attribute. The third model is built after eliminating the next least effective attribute. This process continues until the final model is created with only the highest ranking input attribute. To be sure, this is a brute force technique that does not take into account the interaction between input attributes. However, this one gene at a time approach is common with high-dimensional data.

Attribute ranking is accomplished with the help of the *GainRatioAttributeEval* attribute selection function within RWeka. *GainRatioAttributeEval* evaluates an individual attribute's worth by measuring its gain ratio with respect to the class.

Script 10.4 shows the process and partial output of the statements that prepare the data for our second experiment. We first give *GainRatioAttributeEval* the name of the dataset and the output attribute. *GainRatioAttributeEval* ranks each attribute and places the attribute rankings into sbr.eval. The sort function sorts the attributes by rank. As the ranking is from lowest to highest, we must specify the ranking is to be in descending order. Next, the *select* function found within the *sqldf* package is applied to build a new dataset containing the output attribute and the 30 most individually relevant input attributes. Lastly, the new dataset is used to create the training and test data.

Script 10.4 Ordering the SBRCT Dataset by Attribute Relevance

```
> library(RWeka)
> library(e1071)
> library(sqldf)

> sbr.eval<- GainRatioAttributeEval(Class ~ ., data=sbr.data)
> sbr.eval <-sort(sbr.eval, decreasing=TRUE)
> sbr.eval

# The top 6 features as ranked by GainRatioAttributeEval.
# Your output will show the rank of all 59 attributes.

    1        2        3        4        5        6

X241412 X784224 X796258 X1469292 X767183 X810057

> sbrdata <- sbr.data # sql doesn't like attributes with decimals

# Create a new dataset having the class attribute and the
# first 30 attributes ranked from most to least relevant.

> newsbr.data <- sqldf("select Class,X241412 ,X784224, X796258,
+                      X1469292, X767183 ,X810057 , X183337,
+                      X814260 , X295985 ,X377461 , X298062,
+                      X769657 , X1435862,X629896 , X461425,
+                      X207274 , X325182 ,X812105 , X43733,
+                      X866702 , X244618 ,X296448 , X297392,
+                      X52076  , X841641 ,X624360 , X357031,
+                      X785793 ,X840942  ,X770394
+              from sbrdata")

# Create the training and test data

> newsbr.train <- newsbr.data[1:61,]
> newsbr.test  <- newsbr.data[62:88,]
```

The attribute elimination process is displayed in Script 10.5. The *for loop* shows that model building focuses on the 30 best input attributes. Notice how the *for loop* variable *k* is incorporated to create each modified dataset. A comment symbol has been placed in front of several print statements. Remove comment symbol to learn more about the individual models.

Script 10.5 Attribute Elimination Using the SBRCT Dataset

```
> for (k in 30:2)
+ {
```

```
+ # Build the model and examine the summary stats
+ sbr.model <-svm(Class ~ ., data=newsbr.train)
+ # print(summary(sbr.model))
+
+ # Apply the model to the test data and output the confusion
matrix.
+ newsbr.pred <- predict(sbr.model,newsbr.test)
+
+ # create and print the confusion matrix
+ sbr.perform <- table(newsbr.test$Class,newsbr.
pred,dnn=c("Actual", +"Predicted"))

+ # print(sbr.perform)
+ cat("Number of input attributes=",k,"\n")
+ # Compute and print the % correct.
+ confusionP(sbr.perform)
+
+ # Remove the least predictive individual attribute.
+ # Create the new training and test datasets.
+
+ newsbr.data <- newsbr.data[ ,1:k]
+ newsbr.train <- newsbr.data[1:61,]
+ newsbr.test <- newsbr.data[62:88,

# Partial Output List

Number of input attributes= 30
% Correct= 81.48
Number of input attributes= 28
% Correct= 92.59
Number of input attributes= 19
% Correct= 100
Number of input attributes= 18
% Correct= 100
Number of input attributes= 16
% Correct= 92.59
Number of input attributes= 6
% Correct= 81.48
Number of input attributes= 5
% Correct= 62.96
Number of input attributes= 2
% Correct= 55.56
```

The partial output list shows that classification correctness quickly increases as a result of eliminating the least relevant input attributes. Classification correctness on several models is 100%! Also, after moving from six to five input attributes, the accuracy drops from over 80% to the 60% range. Although these results are very encouraging, the small test set size does not allow us to determine if the differences in test set accuracy are significant. Permutation testing is an appropriate option for obtaining insight into the significance of these differences. Additional experiments with these data are provided in the exercises below.

This concludes our very brief discussion of microarray data mining. If your curiosity has been stimulated beyond the scope of this discussion, a starting point for finding out more can be found at the internet sites listed below. The first link is a software package specific to R.

http://master.bioconductor.org/

https://www.genome.gov/about-genomics

https://www.ebi.ac.uk/arrayexpress/

In addition to the links given above, the Power Point slides available at the book's Web site offer a list of more microarray resources.

10.6 CHAPTER SUMMARY

SVMs offer a unique approach for classifying both linearly separable and nonlinear data. SVMs are able to provide globally optimal problem solutions. Also, because SVMs use the most difficult-to-classify instances as the basis for their predictions, they are less likely to over fit the data.

Even with the use of a kernel function, SVMs cannot compete with the classification times seen with decision trees and several other techniques. However, their accuracy in difficult-to-classify domains makes SVMs a valuable classification tool. Of particular interest is the use of SVMs in finding anomalies in large-sized data.

Added difficulties with SVMs include questions about computational complexity, dimensionality (finding a minimal dimension for achieving linear separation), choosing appropriate attributes, and selecting suitable mapping functions. SVMs are also limited to two class problems with numeric input.

The last issues have been addressed in that we know how categorical input attributes can be transformed to numeric values by simply making each attribute value a binary attribute. Also, the two class limitation has been dealt with in several ways. A common method known as *one-versus-all* builds a classifier model for each class. For n classes, we have n models. An unknown instance is classified with the class of closest association. The *one-vs-one* method builds a classifier for each pair of classes. Given n classes, a total of $n(n-1)/2$ classes are constructed. An unknown instance is classified with the class getting the most votes.

R offers several packages with support vector functions. Our discussion here focused on RWeka's SMO function and the SVM contained in the e1017 package. If you wish to investigate another SVM implementation, the *ksvm* function found within the *kernlab* package is a viable choice.

10.7 KEY TERMS

- *Dot product.* Algebraically, the dot product of two vectors is the sum of the products of their corresponding entries.

- *Gene filtering.* Removing genes that display little variation. For example, genes that have a range less than a fixed value are removed.

- *Hyperplane.* A subspace one dimension less than its ambient space.

- *Maximum margin hyperplane.* The hyperplane showing the greatest separation between two classes.

- *Microarray.* A collection of microscopic DNA spots attached to a solid surface.

- *Microarray data mining.* The process of analyzing the gene expression data obtained from microarray devices.

- *One-vs-one method.* This method is used with models limited to two class problems. When more than two classes are present, this method builds a model for each pair of classes.

- *One-vs-all method.* This method is used with models limited to two class problems. When more than two classes are present, this method builds a model for each class.

- *Permutation testing.* A randomization technique used to help identify false positive correlations between the input attributes and the output. The values of the output attribute are repeatedly randomly assigned to the input instances in order to determine which input attributes are able to significantly differentiate the classes defined by the output attribute.

- *Support vector.* The boundary instances associated with the MMH.

- *Thresholding.* Setting an upper and lower bound value for gene attributes. Attribute values that do not fall within the specified range are discarded.

EXERCISES

Review Questions

1. Differentiate between the following:

   a. One-vs-one and one-vs-all

   b. Linear and nonlinear SVMs.

2. Explain how SMO and *svm* handle categorical data. Read the documentation for each function to obtain help.

3. Read the documentation for SMO to describe the difference between na.action settings na.omit, and na.fail.

Experimenting with R

1. Modify the code in Scripts 10.1 and 10.2 to allow data normalization. For each session, use the resulting equation for the maximum margin hyperplane to determine if the points are correctly classified. What do you conclude?

2. Use Excel to open DNA Lung Cancer.xlsx and read the description for the data. Load DNALungCancer.csv into RStudio. Delete the *record name* column. Create a training set using the first 103 records and a test set using the remaining records. Use *svm* together with the training data to build your classifier. Apply your model to the test data. Write a short summary of your results. Include a confusion matrix as well as the 95% test set confidence interval limits for the accuracy of your model.

3. Repeat the experiment given in Scripts 10.4 and 10.5 but include all 59 attributes in your experiment. Write a report that includes a table showing the classification correctness for the top ten models. The table should also have the number of input attributes used for each model. Include in your report a statement as to whether the individual models tend to make the same incorrect test set classifications.

4. Repeat the experiments in Scripts 10.4 and 10.5 but limit the attribute selection to the 20 least relevant attributes. Your first model is created from the 20 least relevant attributes, your second model is created with 19 attributes, whereby the most irrelevant attribute is removed. This process continues until the final model is created using the best of the 20 least relevant attributes. Write a summary of your findings that includes a critical analysis of how useful the *GainRatioAttribute* function is for determining attribute relevance with this dataset.

5. Repeat Exercise 3, but use RWeka's InfoGainAttributeEval function to evaluate the attributes. Is GainRatioAttributeEval's attribute ordering the same or similar to the order determined by InfoGainAttributeEval?

6. Use the *hist* function to create graphs of the three most predictive and the three least predictive genes within *SBRCTData* as determined by the *GainRationAttributeEval* function. Compare the histogram charts. Can you make any general conclusions?

7. For this exercise, you are to apply SMO to SensorData.csv which has 2212 instances representing three classes. For problems with more than two classes, SMO uses the *one-vs-one* technique described above. The instances contain sensor data collected by a company specializing in aeronautical products.

   a. Use Excel to open *SensorData.xlsx* and read the description of the data.

   b. Import SensorData.csv into Rstudio. Randomize the data. Create a training set with two thirds of the instances in your randomized dataset. Use the remaining instances for the test dataset. Use SMO to build and test a model representing the three classes. Report your results.

c. Create a new dataset having the three most relevant input attributes and the output attribute using an attribute selection technique of your choice. Repeat your experiment.

d. Is there a *significant* difference between the accuracy of the two models?

Computational Questions

1. Use algebra to verify that Equation 10.8 satisfies the condition stated in Equation 10.7.

2. Suppose the star class contains instances (0, 0), (0, 3), and (1, 0). Also suppose the circle class shows instances (4, 2), (4,–1), (5,1), and (4.5,0).

   a. Use simple Euclidean distance to list and verify the support vectors for each class.

   b. Use the method described in this chapter to determine the MMH. Show the system of equations you used to obtain your result.

   c. Use SMO or *svm* with data normalization disabled to verify your result.

Installed Packages

Package Name	Function(s)
base / stats	*cat, for, library, predict, print, sort, summary, table,*
RWeka	*SOM, GainRatioAttributeEval*
e1071	*svm*
sqldf	*sqldf*

## CHAPTER 11

Unsupervised Clustering Techniques

In This Chapter

- Partitioning with the K-Means Algorithm

- Agglomerative Clustering

- Conceptual Clustering

- Expectation Maximization

- Evaluation

IN CHAPTER 8, WE showed you how to perform unsupervised clustering using a neural network approach. In this chapter, we provide an overview of several additional unsupervised clustering techniques. It is not necessary for you to study each technique in detail to obtain an overall understanding of clustering. For this reason, each section of this chapter is self-contained.

Although these techniques are generally considered to be statistical, only the *expectation maximization* (EM) algorithm actually makes limiting assumptions about the nature of the data.

In Section 11.1, we illustrate how the K-Means algorithm partitions instances into disjoint clusters. The focal point of Section 11.2 is agglomerative clustering. Cobweb's incremental hierarchical clustering technique is the topic of Section 11.3. In Section 11.4, we show how the EM algorithm uses classical statistics to perform unsupervised clustering. Section 11.5 is all about unsupervised clustering with R. Let's get started!

11.1 THE K-MEANS ALGORITHM

The K-Means algorithm (Lloyd, 1982) is a simple yet effective statistical clustering technique for numeric data. To help you better understand unsupervised clustering, let's see how the K-Means algorithm partitions a set of data into disjoint clusters.

Here is the algorithm:

1. Choose a value for K, the total number of clusters to be determined.
2. Choose K instances (data points) within the dataset at random. These are the initial cluster centers.
3. Use simple Euclidean distance to assign the remaining instances to their closest cluster center.
4. Use the instances in each cluster to calculate a new mean for each cluster.
5. If the new mean values are identical to the mean values of the previous iteration, the process terminates. Otherwise, use the new means as the cluster centers and repeat steps 3–5.

The first step of the algorithm requires an initial decision about how many clusters we believe to be present in the data. Next, the algorithm randomly selects K data points as initial cluster centers. Each instance is then placed in the cluster to which it is most similar. Similarity can be defined in many ways; however, the similarity measure most often used is simple Euclidean distance.

Once all instances have been placed in their appropriate cluster, the cluster centers are updated by computing the mean of each new cluster. The process of instance classification and cluster center computation continues until an iteration of the algorithm shows no change in the cluster centers. That is, until no instances change cluster assigments in step 3.

11.1.1 An Example Using K-Means

To clarify the process, let's work through a partial example containing two numeric attributes. Although most real datasets contain several attributes, the methodology remains the same regardless of the number of attributes. For our example, we will use the six instances shown in Table 11.1. For simplicity, we name the two attributes x and y, respectively, and map the instances onto an x–y coordinate system. The mapping is shown in Figure 11.1.

As the first step, we must choose a value for K. Let's assume we suspect two distinct clusters. Therefore, we set the value for K at 2. The algorithm chooses two points at random

TABLE 11.1 K-Means Input Values

Instance	X	Y
1	1.0	1.5
2	1.0	4.5
3	2.0	1.5
4	2.0	3.5
5	3.0	2.5
6	5.0	6.0

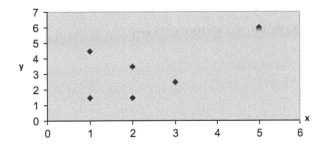

FIGURE 11.1   A coordinate mapping of the data in Table 11.1.

to represent initial cluster centers. Suppose the algorithm selects instance 1 as one cluster center and instance 3 as the second cluster center. The next step is to classify the remaining instances.

Recall the formula for computing the Euclidean distance between point A with coordinates (x_1, y_1) and point B with coordinates (x_2, y_2):

$$\text{Distance}(A - B) = \sqrt{(x_1 - x_2)^2 + (y_1 - y_2)^2}$$

Computations for the first iteration of the algorithm with $C_1 = (1.0, 1.5)$ and $C_2 = (2.0, 1.5)$ are as follows where $C_i - j$ is the Euclidean distance from point C_i to the point represented by instance j in Table 11.1.

Distance $(C_1 - 1) = 0.00$ Distance $(C_2 - 1) = 1.00$

Distance $(C_1 - 2) = 3.00$ Distance $(C_2 - 2) \cong 3.16$

Distance $(C_1 - 3) = 1.00$ Distance $(C_2 - 3) = 0.00$

Distance $(C_1 - 4) \cong 2.24$ Distance $(C_2 - 4) = 2.00$

Distance $(C_1 - 5) \cong 2.24$ Distance $(C_2 - 5) \cong 1.41$

Distance $(C_1 - 6) \cong 6.02$ Distance $(C_2 - 6) \cong 5.41$

After the first iteration of the algorithm, we have the following clustering:

C_1 contains instances 1 and 2.

C_2 contains instances 3, 4, 5, and 6.

The next step is to recompute each cluster center.
For cluster C_1:

$x = (1.0 + 1.0)/2 = 1.0$

$y = (1.5 + 4.5)/2 = 3.0$

For cluster C_2:

$x = (2.0 + 2.0 + 3.0 + 5.0)/4.0 = 3.0$

$y = (1.5 + 3.5 + 2.5 + 6.0)/4.0 = 3.375$

Thus, the new cluster centers are $C_1 = (1.0,3.0)$ and $C_2 = (3.0,3.375)$. As the cluster centers have changed, the algorithm must perform a second iteration.

Computations for the second iteration are:

Distance $(C_1 - 1) = 1.50$ Distance $(C_2 - 1) \cong 2.74$

Distance $(C_1 - 2) = 1.50$ Distance $(C_2 - 2) \cong 2.29$

Distance $(C_1 - 3) \cong 1.80$ Distance $(C_2 - 3) = 2.125$

Distance $(C_1 - 4) \cong 1.12$ Distance $(C_2 - 4) \cong 1.01$

Distance $(C_1 - 5) \cong 2.06$ Distance $(C_2 - 5) = 0.875$

Distance $(C_1 - 6) = 5.00$ Distance $(C_2 - 6) \cong 3.30$

The second iteration results in a modified clustering:

C_1 contains instances 1, 2, and 3.

C_2 contains instances 4, 5, and 6.

Next, we compute the new centers for each cluster.
For cluster C_1:

$x = (1.0 + 1.0 + 2.0)/3.0 \cong 1.33$

$y = (1.5 + 4.5 + 1.5)/3.0 = 2.50$

For cluster C_2:

$x = (2.0 + 3.0 + 5.0)/3.0 \cong 3.33$

$y = (3.5 + 2.5 + 6.0)/3.0 = 4.00$

Once again, this iteration shows a change in the cluster centers. Therefore, the process continues to a third iteration with $C_1 = (1.33,2.50)$ and $C_2 = (3.33,4.00)$. We leave the computations for the third iteration as an exercise.

These computations have little meaning other than to demonstrate the workings of the algorithm. In fact, we may see a different final cluster configuration for each alternative choice of the initial cluster centers. Unfortunately, this is a general problem seen with the

K-Means algorithm. That is, although the algorithm is guaranteed to cluster the instances into a stable state, the stabilization is not guaranteed to be optimal.

An optimal clustering for the K-Means algorithm is frequently defined as a clustering that shows a minimum summation of squared error differences between the instances and their corresponding cluster center. Finding a globally optimal clustering for a given value of K is nearly impossible as we must repeat the algorithm with alternative choices for the initial cluster centers. For even a few hundred data instances, it is not practical to run the K-Means algorithm more than a few times. Instead, the usual practice is to choose a terminating criterion, such as a maximum acceptable squared error value, and execute the K-Means algorithm until we achieve a result that satisfies the termination condition.

Table 11.2 shows three clusterings resulting from repeated application of the K-Means algorithm to the data in Table 11.1. Figure 11.2 displays the most frequently occurring clustering. This clustering is shown as outcome 2 in Table 11.2. Notice that the best clustering, as determined by a minimum squared error value, is outcome 3, where the single instance with coordinates (5,6) forms its own cluster and the remaining instances shape the second cluster.

11.1.2 General Considerations

The K-Means method is easy to understand and implement. However, there are several issues to consider. Specifically,

- The algorithm only works with real-valued data. If we have a categorical attribute in our dataset, we must either discard the attribute or convert the attribute values to numerical equivalents. A common approach is to create one numeric attribute for each categorical attribute value.

TABLE 11.2   Several Applications of the K-Means Algorithm ($K = 2$)

Outcome	Cluster Centers	Cluster Points	Squared Error
1	(2.67,4.67)	2, 4, 6	14.50
	(2.00,1.83)	1, 3, 5	
2	(1.5,1.5)	1, 3	15.94
	(2.75,4.125)	2, 4, 5, 6	
3	(1.8,2.7)	1, 2, 3, 4, 5	9.60
	(5,6)	6	

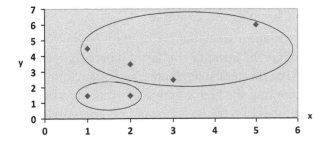

FIGURE 11.2   A K-Means clustering of the data in Table 11.1 ($K = 2$).

- We are required to select a value for the number of clusters to be formed. This is an obvious problem if we make a poor choice. One way to cope with this issue is to run the algorithm several times with alternative values for K. In this way, we are more likely to get a "feel" for how many clusters may be present in the data.

- The K-Means algorithm works best when the clusters that exist in the data are of approximately equal size. This being the case, if an optimal solution is represented by clusters of unequal size, the K-Means algorithm is not likely to find a best solution.

- There is no way to tell which attributes are significant in determining the formed clusters. For this reason, several irrelevant attributes can cause less than optimal results.

- A lack of explanation about the nature of the formed clusters leaves us responsible for much of the interpretation about what has been found. However, we can use supervised machine learning tools to help us gain insight into the nature of the clusters formed by unsupervised clustering algorithms.

Despite these limitations, the K-Means algorithm continues to be a favorite statistical technique.

11.2 AGGLOMERATIVE CLUSTERING

Agglomerative clustering is a favorite unsupervised clustering technique. Unlike the K-Means algorithm, which requires the user to specify the number of clusters to be formed, agglomerative clustering begins by assuming that each data instance represents its own cluster. The steps of the algorithm are as follows:

1. Begin by placing each data instance into a separate partition.
2. Until all instances are part of a single cluster:
   a. Determine the two most similar clusters (smallest distance).
   b. Merge the two clusters chosen in part a into a single cluster.
3. Choose a clustering formed by one of the step 2 iterations as a final result.

Let's see how agglomerative clustering can be applied to the credit card promotion database!

11.2.1 Agglomerative Clustering: An Example

Table 11.3 shows five instances from the credit card promotion database described in Chapter 1. Table 11.4 offers the instance similarity scores computed for the first iteration of algorithm step 2(a). For categorical attributes, an instance-to-instance similarity score is computed by first counting the total number of attribute-value matches between two instances. The total number of attribute comparisons then divides this total, giving the similarity measure. For example, comparing I_1 with I_3, we have four matches and five compares. This gives us the 0.80 value seen in row I_3, column I_1 of Table 11.4.

TABLE 11.3 Five Instances from the Credit Card Promotion Database

Instance	Income Range	Magazine Promotion	Watch Promotion	Life Insurance Promotion	Gender
I_1	40–50 K	Yes	No	No	Male
I_2	25–35 K	Yes	Yes	Yes	Female
I_3	40–50 K	No	No	No	Male
I_4	25–35 K	Yes	Yes	Yes	Male
I_5	50–60 K	Yes	No	Yes	Female

TABLE 11.4 Agglomerative Clustering: First Iteration

	I_1	I_2	I_3	I_4	I_5
I_1	1.00				
I_2	0.20	1.00			
I_3	0.80	0.00	1.00		
I_4	0.40	0.80	0.20	1.00	
I_5	0.40	0.60	0.20	0.40	1.00

Step 2(b) requires a merger of the two most similar clusters formed in the first iteration. As table combinations $I_1 - I_3$ and $I_2 - I_4$ each show the highest similarity score, we can choose to merge I_1 with I_3 or I_2 with I_4. We choose to merge I_1 with I_3. Therefore, after the first iteration of step 2(b), we have three single-instance clusters (I_2, I_4, I_5) and one cluster having two instances (I_1 and I_3).

It is important to note that we could also represent this table as a *dissimilarity matrix* where attribute matches would not count toward the dissimilarity score. If we had chosen dissimilarity, we would merge instances or clusters having the lowest table score.

Next, we need a method to compute cluster-to-cluster similarity scores. Several possibilities exist. For our example, we use the average similarity of all instances involved in a single table computation. That is, to compute a score for merging the cluster containing I_1 and I_3 with the cluster having I_4, we divide 7 attribute-value matches by 15 compares, giving a similarity score of 0.47. This similarity score is seen in row I_4 column $I_1 I_3$ of Table 11.3. Note that the table gives all similarity scores for the second iteration of the algorithm.

In the preceding sequence of actions, we had the choice of merging I_1 with I_3 or I_2 with I_4. When arbitrary choices like this occur, different choices can result in different clusterings. For this reason, it is wise to run several clustering iterations and then choose the best clustering using one of the approaches outlined below.

Table 11.5 tells us that the next iteration of the algorithm will merge I_4 with I_2. Therefore, after the third iteration of step 2, we have three clusters: One cluster containing I_4 and I_2, a second cluster having I_1 and I_3, and the final cluster with I_5 as its only member. The merging of individual clusters continues until all instances are part of a single cluster.

The last step, requiring the choice of a final clustering, is the most difficult. Several statistical as well as heuristic measures can be applied. The following are three simple heuristic techniques that work well in a majority of situations. Specifically,

TABLE 11.5   Agglomerative Clustering: Second Iteration

	$I_1 I_3$	I_2	I_4	I_5
$I_1 I_3$	0.80			
I_2	0.33	1.00		
I_4	0.47	0.80	1.00	
I_5	0.47	0.60	0.40	1.00

1. Invoke the similarity measure used to form the clusters and compare the average within-cluster similarity to the overall similarity of all instances in the dataset. The overall or domain similarity is simply the score seen with the final iteration of the algorithm. In general, if the average of the individual cluster similarity scores shows a higher value than the domain similarity, we have positive evidence that the clustering is useful. As several clusterings of the algorithm may show the desired quality, this technique is best used to eliminate clusterings rather than to choose a final result.

2. As a second approach, we compare the similarity within each cluster to the similarity between each cluster. For example, given three clusters, A, B, and C, we analyze cluster A by computing three scores. One score is the within-class similarity of the instances in cluster A. The second score is obtained by computing the similarity score seen when all the instances of cluster A are combined with the instances of B. The third score is the similarity value obtained by grouping the instances of cluster C with those of cluster A. We expect to see the highest scores for within-class similarity computations. As with the first technique, several clusterings of the algorithm may show the desired quality. Therefore, the technique is also best used to eliminate clusterings rather than to choose a final result.

3. One useful measure to be used in conjunction with the previous techniques is to examine the rule sets generated by each saved clustering. For example, let's assume ten iterations of the algorithm are initially performed. Next, suppose five of the ten clusterings are eliminated by using one of the previous techniques. To apply the current method, we present each clustering to a rule generator and examine the rules created from the individual clusters. The clustering showing a best set of rules, based on precision and coverage, is chosen as the final result.

Provided that certain assumptions can be made about the data, a statistical analysis may also be applied to help determine which of the clusterings gives a best result. One common statistical test used to select a best partitioning is the *Bayesian Information Criterion* also known as the *BIC*. The BIC requires that the clusters be normally distributed. The BIC gives the odds for one model against another model assuming that neither model is initially favored (Dasgupta and Raftery, 1998).

11.2.2 General Considerations

Agglomerative clustering creates a hierarchy of clusterings by iteratively merging pairs of clusters. Although we have limited our discussion here, several procedures for computing cluster similarity scores and merging clusters exist. When data are real-valued, defining a measure of instance similarity can be a challenge. One common approach is to use simple Euclidean distance.

A widespread application of agglomerative clustering is its use as a prelude to other clustering techniques. For example, the first iteration of the K-Means algorithm requires a choice for initial cluster means. In the usual case, the choice is made in a random or arbitrary manner. However, the initial selection can have a marked effect on goodness of the final clustering. Therefore, to increase our chance of obtaining a best final clustering, we first apply agglomerative clustering to create the same number of clusters as that chosen for the K-Means algorithm. Next, we compute the cluster means resulting from the agglomerative technique and use the mean scores as the initial choice for the first K-Means clustering (Mukherjee et al., 1998).

11.3 CONCEPTUAL CLUSTERING

Conceptual clustering is an unsupervised clustering technique that incorporates incremental learning to form a hierarchy of concepts. The concept hierarchy takes the form of a tree structure where the root node represents the highest level of concept generalization. Therefore, the root node contains summary information for all domain instances. Of particular interest are the basic-level nodes of the tree. The basic-level nodes are interesting in terms of human appeal and understanding. An appropriate measure of cluster quality forms these basic-level nodes at the first or second level of the concept tree. The following is a standard conceptual clustering algorithm:

1. Create a cluster with the first instance as its only member.
2. For each remaining instance, take one of two actions at each level of the tree:
   a. Place the new instance into an existing cluster.
   b. Create a new concept cluster having the presented instance as its only member.

The algorithm clearly shows the incremental nature of the clustering process. That is, each instance is presented to the existing concept hierarchy in a sequential manner. Next, at each level of the hierarchy, an evaluation function is used to make a decision about whether to include the instance in an existing cluster or to create a new cluster with the new instance as its only member. In the next section, we describe the evaluation function used by a well-known probability-based conceptual clustering system.

11.3.1 Measuring Category Utility

Cobweb (Fisher, 1987) is a conceptual clustering model that stores knowledge in a concept hierarchy. Cobweb accepts instances in attribute-value format where attribute values must be categorical. Cobweb's evaluation function has been shown to consistently determine psychologically preferred (basic) levels in human classification hierarchies. The evaluation function is a generalization of a measure known as *category utility*. The category utility function measures the gain in the "expected number" of correct attribute-value predictions for a specific object if it were placed within a given category.

The formula for category utility includes three probabilities. One measure is the conditional probability of attribute A_i having value V_{ij} given membership in class C_k, denoted as $P(A_i = V_{ij} \mid C_k)$. This is the formal definition of attribute-value *predictability*. If the value of $P(A_i = V_{ij} \mid C_k) = 1$, we can be certain that each instance of class C_k will always have V_{ij} as the value for attribute A_i. Attribute A_i having value V_{ij} is said to be a *necessary* condition for defining class C_k. The second probability, $P(C_k \mid A_i = V_{ij})$, is the conditional probability that an instance is in class C_k given that attribute A_i has value V_{ij}. This is the definition of attribute-value *predictiveness*. If the value of $P(C_k \mid A_i = V_{ij})$ is 1, we know that if A_i has value V_{ij}, the class containing this attribute-value pair must be C_k. Attribute A_i having value V_{ij} is said to be a *sufficient* condition for defining class C_k. Given these definitions, we see that attribute-value predictability is a within-class measure and attribute-value predictiveness is a between-class measure.

These three probability measures are combined and summed across all values of i, j, and k to describe a heuristic measure of partition quality. Specifically,

$$\sum_K \sum_i \sum_j P(A_i = V_{ij}) \, P(C_K \mid A_i = V_{ij}) \, P(A_i = V_{ij} \mid C_K) \tag{11.1}$$

The probability $P(A_i = V_{ij})$ allows attribute values that are seen frequently to play a more important part in measuring partition quality. Using the expression for partition quality, category utility is defined by the formula:

$$\frac{\sum_{k=1}^{K} P(C_K) \sum_i \sum_j P(A_i = V_{ij} \mid C_k)^2 - \sum_i \sum_j P(A_i = V_{ij})^2}{K} \tag{11.2}$$

The first numerator term is the previously described partition-quality expression stated in an alternative form through the application of Bayes rule. The second term represents the probability of correctly guessing attribute values without any category knowledge. The division by k (total number of classes) allows Cobweb to consider variations in the total number of formed clusters.

11.3.2 Conceptual Clustering: An Example

To illustrate the process used by Cobweb to build a concept hierarchy, consider the hierarchy shown in Figure 11.3 created by Cobweb when presented with the instances in Table 11.6. Let's suppose we have a new instance to be placed in the hierarchy. The instance enters the hierarchy at root node N, and N's statistics are updated to reflect the addition of the new instance. As the instance enters the second level of the hierarchy, Cobweb's evaluation function chooses one of four actions. If the new instance is similar enough to one of N_1, N_2, or N_4, the instance is incorporated into the preferred node and the instance proceeds to the second level of the hierarchy through the chosen path. As a second choice, the evaluation function can decide that the new instance is unique enough to merit the creation of a new first-level concept node. This being the case, the instance becomes a first-level concept node and the classification process terminates.

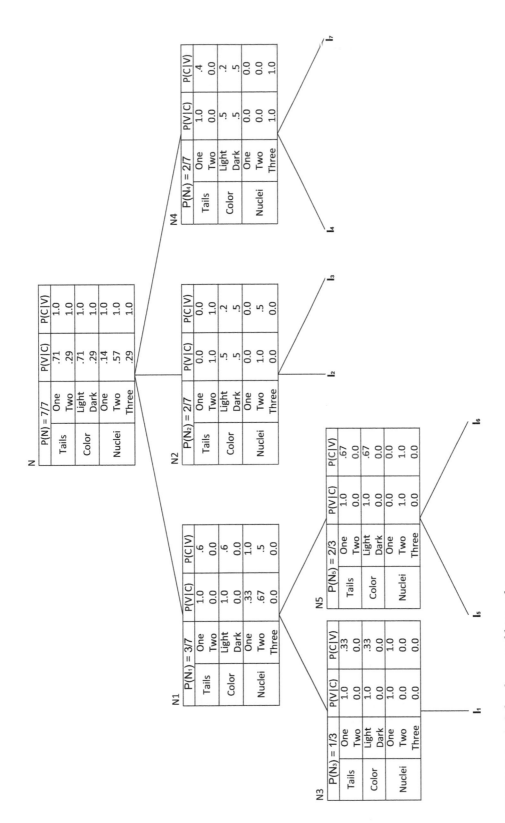

FIGURE 11.3   A Cobweb-created hierarchy.

TABLE 11.6   Data for Conceptual Clustering

	Tails	Color	Nuclei
I_1	One	Light	One
I_2	Two	Light	Two
I_3	Two	Dark	Two
I_4	One	Dark	Three
I_5	One	Light	Two
I_6	One	Light	Two
I_7	One	Light	Three

Cobweb allows two other choices. In one case, the system considers merging the two best-scoring nodes into a single node. The last possibility actually removes the best-scoring node from the hierarchy. These final two choices are to help modify nonoptimal hierarchies that can result from skewed instance presentation—several highly similar instances are presented in sequence. If either choice is made, the merge or delete operation is processed and the new instance is once again presented for classification to the modified hierarchy. This procedure continues at each level of the concept tree until the new instance becomes a terminal node.

As a final point, notice that all predictiveness and predictability scores are computed with respect to the parent node. For example, the predictiveness score of 1.0 for *nuclei = two* found in N_3 reflects the fact that all instances incorporated into N_2 with *nuclei = two* have followed the path from N_2 to N_3.

11.3.3 General Considerations

Although we have limited our discussion to Cobweb, several other conceptual clustering systems have been developed. Classit (Gennari et al., 1989) is a conceptual clustering system that uses an extension of Cobweb's basic algorithm to build a concept hierarchy and classify instances. The two models are very similar with the exception that Classit's evaluation function is an equivalent transformation of Cobweb's category utility for real-valued attributes. Therefore, individual concept nodes store attribute mean and standard deviation scores rather than attribute-value probabilities.

Like Cobweb, Classit is particularly appealing as its evaluation function has been shown to consistently determine psychologically preferred levels in human classification hierarchies. In addition, both Cobweb and Classit lend themselves well to explaining their behavior because each tree node contains a complete concept description at some level of abstraction.

Conceptual clustering systems also have several shortcomings. A major problem with conceptual clustering systems is that instance ordering can have a marked impact on the results of the clustering. A nonrepresentative ordering of the instances can lead to a less than optimal clustering. Clustering systems such as Cobweb and Classit use special operations to merge and split clusters in an attempt to overcome this problem. However, the results of these techniques have not been shown to be successful in all cases.

11.4 EXPECTATION MAXIMIZATION

The EM algorithm (Dempster et al., 1977) is a statistical technique that makes use of the finite Gaussian mixtures model. A *mixture* is a set of n probability distributions where each distribution represents a cluster. The mixtures model assigns each individual data instance a probability (rather than a specific cluster) that it would have a certain set of attribute values given it was a member of a specified cluster. The mixtures model assumes all attributes to be independent random variables.

The EM algorithm is similar to the K-Means procedure in that a set of parameters are recomputed until a desired convergence value is achieved. In the simplest case, $n = 2$, the probability distributions are assumed to be normal, and data instances consist of a single real-valued attribute. Although the algorithm can be applied to a datasets having any number of real-valued attributes, we limit our discussion here to this simplest case and provide a general example in the next section. Using the two-class, one-attribute scenario, the job of the algorithm is to determine the value of five parameters. Specifically,

- The mean and standard deviation for cluster 1

- The mean and standard deviation for cluster 2

- The sampling probability P for cluster 1—the number of instances in cluster 1 divided by the total number of domain instances. The probability for cluster 2 is therefore $(1 - P)$.

The general procedure used by EM is as follows:

1. Guess initial values for the five parameters given above
2. Until a specified termination criterion is achieved:
   a. Use the probability density function for normal distributions (Equation 5.16) to compute the cluster probability for each instance. In the two-cluster case, we have two probability distribution formulas, each with differing mean and standard deviation values.
   b. Use the probability scores assigned to each instance in step 2(a) to reestimate the five parameters.

The algorithm terminates when a formula that measures cluster quality no longer shows significant increases. One measure of cluster quality is the likelihood that the data came from the dataset determined by the clustering. The likelihood computation is straightforward. For each data instance, first sum the conditional probabilities that the instance belongs in each of the two clusters. In this way, each instance has an associated summed probability score. Next, multiply the summed probability scores to compute the likelihood value. Higher likelihood scores represent more optimal clusterings. EM's solid statistical foundation as well as its similarity to the K-Means algorithm makes it one of the most referenced clustering algorithms.

EM is guaranteed to converge to a maximum likelihood score. However, the maximum may not be global. For this reason, several applications of the algorithm may be necessary to achieve a best result. As initial mean and standard deviation scores selected by the algorithm affect the final result, an alternative technique such as agglomerative clustering is often initially applied. EM then uses the mean and standard deviation values for the clusters determined by the preliminary technique as initial parameter settings.

A lack of explanation about what has been discovered is a problem with EM as it is with many clustering systems. For this reason, our suggested methodology of using a supervised model to analyze the results of an unsupervised clustering is often appropriate. The next section offers clustering experiments that employ K-Means partition clustering and hierarchical agglomerative clustering.

11.5 UNSUPERVISED CLUSTERING WITH R

Supervised learning and unsupervised clustering complement one another in that each approach can be applied to evaluate the opposite strategy. Evaluating an unsupervised clustering with supervised learning is particularly appealing as the process is independent of the algorithm used to cluster the data.

11.5.1 Supervised Learning for Cluster Evaluation

For our first example, we use supervised learning to help interpret the meaning of the clusters obtained by applying the K-Means algorithm to a subset of the gamma-ray burst dataset described in Chapter 5. Recall this data has been preprocessed with a logarithmic normalization. Here is the procedure:

1. Present the data to the clustering algorithm.

2. Designate each formed cluster as a class.

3. Choose a supervised learning algorithm with explanation capabilities to build a model of the classes.

4. Use the model created in (3) to help explain and analyze the formed clusters.

Script 11.1 displays the steps along with edited output for this experiment. To simplify our explanation, all excepting *t50* and *hr32* have been eliminated from the original data.

The script tells us that the preprocessed data is presented to the *kmeans* function with instructions to form three clusters. The component *grb.km$size* displays the number of instances in each cluster, *grb.km$centers* gives us the individual cluster centers, and *grb.km$cluster* provides a list of the individual cluster number associated with each instance.

Script 11.1 A Decision Tree to Explain a Clustering of Gamma-Ray Burst Data

```
> # Create a dataset with T50 and HR32
> T50Hr32.data<- Grb4u[, c(4,6)]
> #head(T50Hr32.data)
> set.seed(1000)

> # Cluster the data
> grb.km <- kmeans(T50Hr32.data,centers = 3)

> grb.km$size

  433 453 293

> # show centers
> round(grb.km$centers,3)

    hr32     t50
1  0.405   0.551
2  0.423   1.414
3  0.735  -0.839

> # Plot the individual clusters
> plot(T50Hr32.data[grb.km$cluster==1,],col="red",ylim=c(-3,3))
> points(T50Hr32.data[grb.km$cluster==2,],col="blue")
> points(T50Hr32.data[grb.km$cluster==3,],col="green")

> class <- as.factor(grb.km$cluster)
> T50Hr32.data2<- cbind(class,T50Hr32.data)
> library(rpart)
> rpart.model <- rpart(class ~ ., data = T50Hr32.data2,
+ method ='class', control=rpart.control(minsplit = 2), model=TRUE)
> # Plot the tree
> library(rpart.plot)
> rpart.plot(rpart.model
```

Three additional components—not displayed in the script—provide the means for computing an often cited measure of cluster quality given by dividing between sum of squares (*grb. km$betweenss*) by total sum of squares (*grb.km$totss*) where *betweenss = totalss – withinss*. The closer this value is to 1.0, the more confidence we have that the clusters differentiate the data in a meaningful way. Computing this value gives us 0.81. This value lends support to the hypothesis that the clusters capture the structure of the data in a meaningful way. You can obtain a list of all components available with the clustering by typing *grb.km*.

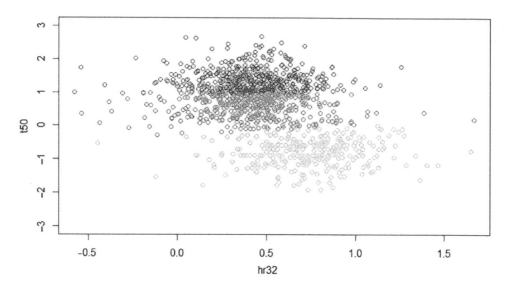

FIGURE 11.4   A K-Means clustering of gamma-ray burst data.

Upon examining the cluster sizes and centers, we see three well-defined clusters. Cluster 1 represents soft-short bursts, cluster 2 characterizes soft-long bursts, and cluster 3 holds very hard-short bursts. Figure 11.4 graphically depicts each cluster. The *points* function allows us to add the clusters to the graph one at a time. Setting *ylim* is necessary for all three clusters to appear in the graph. We see several instances as possible outliers.

Next, the list of cluster numbers is converted to the factor variable *class* which is then bound to the original data. Finally, *rpart* uses the clusters to create the decision tree shown in Figure 11.5.

The decision tree tells us that the clustering constructs a single partition of long bursts (t50≥0.99). The remaining 62% of the instances are again divided by T50. Burst hardness does not play a role in the construction of the tree. However, there is a marked difference in hardness between clusters 2 and 3 with cluster 3 representing the shortest and hardest bursts. For our next experiment, we reexamine how unsupervised clustering can be used for attribute evaluation.

11.5.2 Unsupervised Clustering for Attribute Evaluation

In Chapter 8, we applied unsupervised neural net clustering to a 768 instance dataset where 268 of the individual instances represented females who tested positive for diabetes. Our goal was to evaluate the relevance of the input attributes for supervised learning. Our initial conclusion was that the input attributes were not able to clearly represent the output attribute. Here we revisit this dataset by applying the K-Means algorithm to help verify or refute our previous conclusion.

Script 11.2 presents the steps and edited output. Several items are of interest. You will first notice the output attribute is removed from the data after which the data is normalized. The clusters to classes model accuracy (67.45%) is in line with the Kohonen clustering

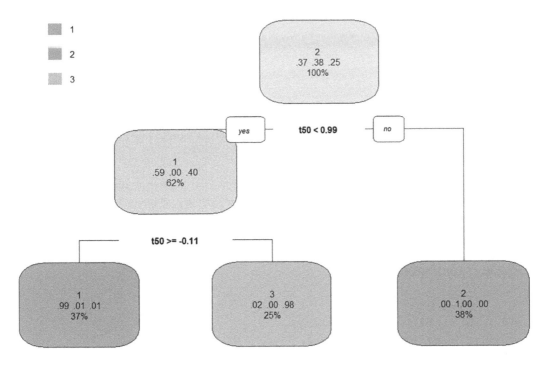

FIGURE 11.5   Decision tree created from a gamma-ray burst clustering.

with these same data (Chapter 8, Section 8.7). The *aggregate* function is invoked to group the instances by their corresponding cluster number. The function then applies the *mean* over the aggregated data. The unedited script also applies *sd* to the aggregated values.

Script 11.2 Clustering Diabetes Data

```
> # Scale the data and set the seed.
> dia.data <- scale(Diabetes[,-9])
> set.seed(100)

> # Perform the K-Means clustering
> dia.km <- kmeans(dia.data,centers = 2)

> # Construct the confusion matrix
cluster.as.char <- ifelse(dia.km$cluster==2,
+ "tested_positive","tested_negative")
> con <-table(Diabetes$Diabetes,cluster.as.char)
> con
```

	1	2
	cluster.as.char	
	tested_negative	tested_positive
"tested_negative"	374	126

```
  "tested_positive"                124              144
> confusionP(con)
  Correct= 518
Incorrect= 250
Accuracy = 67.45 %

> # Convert means to pre-scaled values
> round(aggregate(Diabetes[-9],by =list(dia.km$cluster),FUN =
mean),2)

Group.1 Pregnancies PG.Concent  D.BP  Tri.F Ser.In  BMI  DP.F  Age
     1      2.08       113.70   64.81 21.87 83.67  31.51 0.47 26.48
     2      7.09       134.16   77.03 18.07 72.66  32.89 0.47 45.72

> # Total sum of squares
> round(dia.km$totss,2)
  6136

> # total within sum of squares
> round(dia.km$tot.withinss,2)
  5122.13

> # betweenss = totalss - totalwithinss
> round(dia.km$betweenss,2)
  1013.87

> # betweenss / totalss
> round(dia.km$betweenss / dia.km$totss,2)
  0.17
```

Of primary importance is the value *betweenss /totalss*. The rounded value 0.17 strongly indicates that the clusters do not reflect measurable structure within the attributes. This supports our previous position that the set of input attributes do not differentiate the two classes.

Although this approach to attribute evaluation is straightforward, depending on the clustering technique it may take several iterations before an evaluation can be made. For example, the clusters formed by an application of K-Means are highly affected by the initial selection of cluster means. For a single iteration, the algorithm is likely to experience a less than optimal convergence. With a nonoptimal convergence, a single cluster could contain a mixture of instances from several classes. This would in turn give us a false impression about the efficacy of the domain for supervised learning.

As the unsupervised evaluation is based solely on the training data, a quality clustering is no guarantee of acceptable performance on test set instances. For this reason, the technique complements other evaluation methods and is most valuable for identifying a rationale for supervised model failure. As a general rule, we see an inverse relationship between the value of this approach and the total number of predefined classes contained in the training data. It's time to move on up to hierarchical clustering!

11.5.3 Agglomerative Clustering: A Simple Example

Script 11.3 shows the steps and edited output resulting from applying agglomerative clustering to the data in Table 11.1. The *dist* function creates the *dissimilarity matrix* needed by *hclust* to cluster the data. The matrix—displayed in the script—contains 15 values where each value is the Euclidean distance between one pair of data points. The call to *hclust* shows we are using the *complete linkage* method to define the distance between two clusters. This method defines the distance between two clusters as the greatest distance between a point in one cluster and a point in the second cluster. Other common options include *average linkage* and *single linkage*. *Average linkage* computes the average distance between each point in one cluster and each point in the second cluster. *Single linkage* defines distance as the shortest distance between a point in one cluster and a point in the second cluster. In Script 11.5, we use a technique that makes it possible to cluster datasets having categorical and mixed data types.

Figure 11.6 displays the dendrogram (tree structure) for the clustering. Following the structure from the bottom, we first see each point as its own cluster. Next, points *1* and *3* combine into a single cluster. Further, points *2* and *4* combine to form another cluster. The 5th data point is then added to cluster *1–3*. After this, cluster *1–3–5* combines with cluster *2–4*. Lastly, *6* is added forming the single top-level cluster. You will notice that isolating the 6th point until the final merge corresponds to the third outcome of the K-Means clustering described in Table 11.2.

Script 11.3 Agglomerative Clustering of the Data in Table 11.1

```
> x1 <- c(1,1,2,2,3,5)
> x2 <- c(1.5,4.5,1.5,3.5,2.5,6)
> table11.1.data <- data.frame(x1,x2)

> # Compute and print the dissimilarity matrix.
> ds<-dist(table11.1.data, method='euclidean')

> round(ds,3)

          1     2     3     4     5
2 3.000
3 1.000 3.162
4 2.236 1.414 2.000
5 2.236 2.828 1.414 1.414
6 6.021 4.272 5.408 3.905 4.031

> # Cluster and plot the data.
> my.agg<- hclust(ds,method='complete')
> plot(my.agg)
```

Cluster Dendrogram

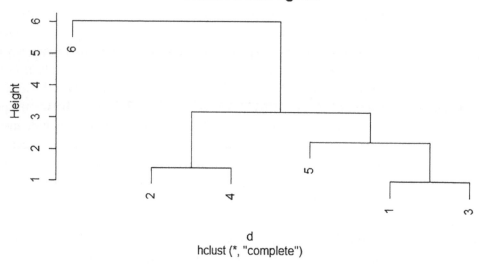

FIGURE 11.6   An agglomerative clustering of the data in Table 11.1.

11.5.4 Agglomerative Clustering of Gamma-Ray Burst Data

Most clustering techniques require the user to make an initial decision about the correct number of clusters in the data. For this example, we employ the gamma-ray burst dataset to illustrate an automated method for making this choice. Script 11.4 displays the process together with edited output. The first order of business is to install package *NbClust*. Second, before looking at Script 11.4, we examine attribute correlations. Specifically,

```
> x<- Grb4u[-1]
> round(cor(x),3)
```

```
          p256      fl    hr32   hr321     t50     t90
p256     1.000   0.602   0.170   0.181   0.013   0.073
fl       0.602   1.000  -0.030  -0.042   0.643   0.683
hr32     0.170  -0.030   1.000   0.959  -0.387  -0.391
hr321    0.181  -0.042   0.959   1.000  -0.407  -0.411
t50      0.013   0.643  -0.387  -0.407   1.000   0.975
t90      0.073   0.683  -0.391  -0.411   0.975   1.000
```

Script 11.4 Agglomerative Clustering of Gamma-Ray Burst Data

```
> # Preprocess the data & determine the maximal no. of clusters.
> library(NbClust)
> grb.data <- Grb4u[,c(-1,-2,-5,-6)]
> numC <- NbClust(grb.data,min.nc=2,max.nc=5,method='average')
               ***** Conclusion *****
```

```
* According to the majority rule, the best number of clusters is 3

> table(numC$Best.n[1,])
 0  2  3  4  5
 2  4 17  1  1

> barplot(table(numC$Best.n[1,]),xlab="No. of
Clusters",ylab="Frequency")

> d<-dist(grb.data, method='euclidean')
> my.agg<- hclust(d,method='complete')

> plot(my.agg)

> clusters <-cutree(my.agg,numC)
> table(clusters)

  clusters
  1   2   3

 622 230 327
> round(aggregate(grb.data, by=list(cluster=clusters),mean),3)

  cluster      fl   hr32     t90
       1  -5.589  0.352   1.354
       2  -4.552  0.552   1.767
       3  -6.367  0.724  -0.366

> round(aggregate(grb.data, by=list(cluster=clusters),sd),3)

  cluster     fl  hr32   t90
       1  0.404 0.239 0.470
       2  0.419 0.143 0.363
       3  0.578 0.316 0.490
```

Recall that the dataset contains one pair of attributes for each of burst length, brightness, and hardness. The correlation table tells us to eliminate one from each pair of: t50-t90, fl- p256, and hr32-hr321. Script 11.4 shows we chose to eliminate t50, p256, and hr321 in addition to the column of unique burst numbers. Next, *NbClust* uses a majority rules approach based on 26 criteria to determine the best number of clusters. The bar plot in Figure 11.7 summarizes the results of the call to *NbClust* telling us that three clusters represent our best option for a final solution. You can see all 26 measures used by *NbClust* by typing numC into the console. Although we first use *NbClust* here, it is independent of the clustering algorithm.

The dendrogram resulting from the clustering is displayed in Figure 11.8. The *cut-Tree* function cuts the dendrogram into three clusters as specified and outputs the

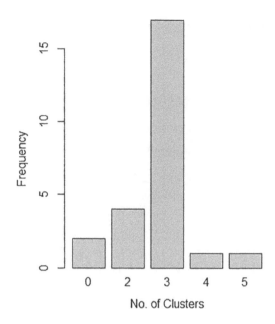

FIGURE 11.7   Determining the number of clusters.

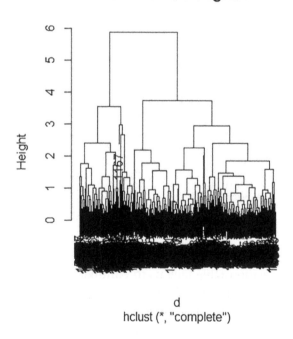

FIGURE 11.8   An agglomerative clustering of gamma-ray burst data.

number of instances in each cluster. Lastly, the *aggregate* function displays within-class mean and standard deviation values. The attribute means can give us a better understanding of the clusters provided they are standardized. One simple heuristic measure of attribute relevance is to divide the absolute value of the difference between attribute means by the global standard deviation for the attribute. Numeric attributes with lower relevance values (usually less than 0.25) will likely be of little value in differentiating one cluster from another.

To illustrate this approach, we can easily obtain global standard deviation scores with the help of *sapply*. For our example we have:

```
> sapply(grb.data,sd)

fl        hr32        t90
0.7684309 0.2964524 0.9515629
```

Having the *sd* values above, let's compare mean values for *t90* between clusters *1* and *2*:

```
> abs(1.354 - 1.767)/.9515629
[1] 0.4340228
```

The same computation for *t90* between clusters *1* and *3* gives us:

```
> abs(-0.366 - 1.767)/.9515629
[1] 2.241575
```

For *fl* and clusters *1* and *2* we obtain:

```
> abs(-5.589- -4.552)/0.7684309
[1] 1.349503
```

These computations indicate that clusters *1* and *3* are differentiated by *t90*, whereas clusters *1* and *2* show a meaningful difference in average burst brightness. In general, the means and standard deviations describe one cluster of longer, brighter, softer bursts; a second cluster of longer, duller, softer bursts; and a third cluster of shorter, harder bursts. Several end-of-chapter exercises ask you to continue exploring the inherent structure of gamma-ray burst data.

11.5.5 Agglomerative Clustering of Cardiology Patient Data

Script 11.5 illustrates how agglomerative clustering can be applied to datasets with categorical and/or mixed data types. To accomplish this, we must install the *cluster package* and use the *daisy* function together with metric = *gower* to create the dissimilarity matrix. The *gower* metric is able to compute similar values for mixed and strictly categorical datasets. If all attributes are numeric, the dissimilarity measure defaults to Euclidean distance.

If one or more categorical variables exist, dissimilarity between two instances is given as the weighted average of the contributions of each attribute. The contribution of a numeric attribute is given as the absolute difference between the values in both instances divided by the range of the numeric attribute. For categorical attributes, the contribution is 0 if both values are identical, otherwise 1. Missing attribute values within an instance are simply not

Cluster Dendrogram

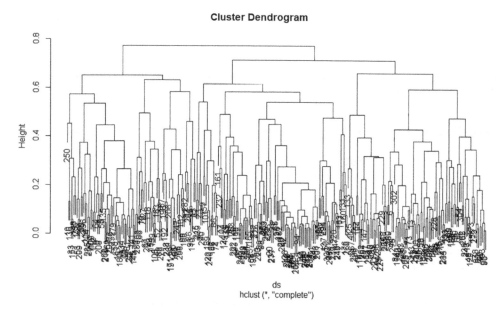

FIGURE 11.9   Agglomerative clustering of cardiology patient data.

included in the dissimilarities involving the instance. You can read about the specifics of this metric by examining the *daisy* documentation within the cluster package.

The dendrogram in Figure 11.9 resulting from the call to *hclust* clearly shows a two cluster split within the data. The creation of the confusion matrix is somewhat problematic as we must interchange the cluster numbers prior to applying the *table* function. This is the case as the first data instance belongs to the *sick* class so the cluster representing those who have had a heart attack will likely find their home in cluster *1*. However, *str(CardiologyMixed)* tells us that factor *1* is associated with the healthy class! One way to resolve this issue is to apply the *for* loop shown in the script where the *1's* and *2's* are interchanged. A better alternative is to place typical instances from each class at the head of the dataset with the first instance matching the class corresponding to factor value *1*. This rule is easily generalized when more than two classes exist in the data. Finally, the clusters to classes confusion matrix with a 76.57% accuracy lends support for the ability of the input attributes to define the structure within the data. You will learn more about clustering datasets with mixed types in the end-of-chapter exercises.

Script 11.5 Agglomerative Clustering of Cardiology Patient Data

```
> library(cluster)
> # extract the class variable from the data
> card.data <- CardiologyMixed[-14]

> # Compute the dissimilarity matrix using the daisy function
```

```
> ds<-daisy(card.data, metric='gower')

> # Cluster and plot the data.
> my.agg<- hclust(ds,method='complete')
> plot(my.agg)

> clusters <-cutree(my.agg,2)
> # The cluster number of each instance (not printed)
> # clusters

> # Modify cluster numbers to correspond with the class factors

> for(i in 1:nrow(CardiologyMixed))
+ {
+    if(clusters[i] ==1)
+       clusters[i]<- 2
+    else
+       clusters[i]<- 1
+ }
> cluster.to.char ifelse(clusters==2,'Sick','Healthy')
> my.conf<- table(cluster.to.char,
+                  CardiologyMixed$class,dnn=c("Actual","Predicted"))
> my.conf
           Predicted
Actual      Healthy Sick
   Healthy      152   58
   Sick          13   80

> confusionP(my.conf)
  Correct= 232
Incorrect= 71
Accuracy = 76.57 %
```

11.5.6 Agglomerative Clustering of Credit Screening Data

For our final example, we applied *hclust* to the *creditScreening* dataset. Recall that 5% of the instances have at least one missing attribute value. As the input data is both categorical and numeric, we used the *daisy* function to create the dissimilarity matrix. We also called the *GainRatioAttributeEval* function to obtain a measure of individual attribute relevance. Script 11.6 shows the steps and edited output for our experiment.

The clusters to classes confusion matrix shows a 56.23% accuracy when all 15 input attributes are used for the clustering. Two hundred ninety-two credit card applications that should have been accepted have been rejected! This, together with the relevance values given by *GainRatioAttributeEval*, indicates that many of the input attributes are irrelevant and negatively influence the outcome. This is easily verified by working through end-of-chapter Exercise 11. Interestingly, you will find that removing all input attributes excepting *nine* will increase classification accuracy by more than 25%!

Script 11.6 Agglomerative Clustering of Credit Screening Data

```
> library(cluster)
> library(RWeka)
> # preprocess the data
> round(sort(GainRatioAttributeEval(class ~ . , data =
+ creditScreening)),3)

   twelve    one  thirteen seven   two    six   four  five  fourteen
   0.001   0.003    0.019  0.029  0.029 0.030 0.033 0.033  0.042
   three   eight   fifteen eleven ten   nine
   0.046   0.114    0.143  0.152 0.161 0.452

> # extract the class variable from the data
> card.data <- creditScreening[-16]

> # Compute the dissimilarity matrix using the daisy function
> ds<-daisy(card.data, metric='gower')

> # Cluster the data.
> my.agg<- hclust(ds,method='complete')
> plot(my.agg)

> clusters <-cutree(my.agg,2)

> # Class is of type 'factor' with 1 = "-" and 2 = "+".
> # As the first instance belongs to the "+" class, we must
> # change the 1's to 2's and vise versa.

> for(i in 1:nrow(card.data))
+ {
+   if(clusters[i] ==1)
+     clusters[i]<- 2
+   else
+     clusters[i]<- 1
+   }
> my.conf<- table(clusters,creditScreening$class,dnn=c("Actual",
+ "Predicted"))
> my.conf

      Predicted
Actual    -    +
```

```
       1   91   10
       2  292  297

> confusionP(my.conf)
  Correct= 388
Incorrect= 302
Accuracy = 56.23 %
```

11.6 CHAPTER SUMMARY

The K-Means algorithm partitons data into a predetermined number of clusters. The algorithm begins by randomly choosing one data point to represent each cluster. Each data instance is then placed in the cluster to which it is most similar. New cluster centers are computed and the process continues until the cluster centers do not change. The K-Means algorithm is easy to implement and understand. However, the algorithm is not guaranteed to converge to a globally optimal solution, lacks the ability to explain what has been found, and is unable to tell which attributes are significant in determining the formed clusters. Despite these limitations, the K-Means algorithm is among the most widely used clustering techniques.

Agglomerative clustering is a favorite hierarchical clustering technique. Agglomerative clustering begins by assuming that each data instance represents its own cluster. Each iteration of the algorithm merges the most similar pair of clusters. The final iteration sees all dataset items contained in a single cluster. Several options for computing instance and cluster similarity scores and cluster merging procedures exist. Also, when the data to be clustered is real-valued, defining a measure of instance similarity can be a challenge. One common approach is to use simple Euclidean distance. A widespread application of agglomerative clustering is its use as a prelude to other clustering techniques.

Conceptual clustering is an unsupervised technique that incorporates incremental learning to form a hierarchy of concepts. The concept hierarchy takes the form of a tree structure where the root node represents the highest level of concept generalization. Conceptual clustering systems are particularly appealing because the trees they form have been shown to consistently determine psychologically preferred levels in human classification hierarchies. Also, conceptual clustering systems lend themselves well to explaining their behavior. A major problem with conceptual clustering systems is that instance ordering can have a marked impact on the results of the clustering. A nonrepresentative ordering of data instances can lead to a less than optimal clustering.

The EM algorithm is a statistical technique that makes use of the finite Gaussian mixtures model. The mixtures model assigns each individual data instance a probability that it would have a certain set of attribute values given it was a member of a specified cluster. The model assumes all attributes to be independent random variables. The EM algorithm is similar to the K-Means procedure in that a set of parameters are recomputed until a desired convergence value is achieved. A lack of explanation about what has been discovered is a problem with EM as it is with many clustering systems. Our methodology of using a supervised model to analyze the results of an unsupervised clustering is one technique to help explain the results of an EM clustering.

Unsupervised clustering techniques often support their own internal evaluation criteria. As many unsupervised techniques offer minimal explanation about the nature of the formed clusters, the evaluation of an unsupervised clustering should include an explanation about what has been discovered. Supervised learning can help explain and evaluate the results of an unsupervised clustering. Another effective evaluative procedure is to perform a between-cluster, attribute-value comparison to determine if the instances contained within the alternative clusters differ significantly.

11.7 KEY TERMS

- *Agglomerative clustering.* An unsupervised technique where each data instance initially represents its own cluster. Successive iterations of the algorithm merge pairs of highly similar clusters until all instances become members of a single cluster. In the last step, a decision is made about which clustering is a best final result.

- *Average linkage.* This method is used with hierarchical clustering and defines distance as the average distance between each point in one cluster and each point in the second cluster.

- *Basic-level nodes.* The nodes in a concept hierarchy that represent concepts easily identified by humans.

- *Bayesian Information Criterion (BIC).* The BIC gives the posterior odds for one data mining model against another model assuming neither model is favored initially.

- *Category utility.* An unsupervised evaluation function that measures the gain in the "expected number" of correct attribute-value predictions for a specific object if it were placed within a given category or cluster.

- *Complete linkage.* This method defines the distance between two clusters as the greatest distance between a point in one cluster and a point in the second cluster.

- *Concept hierarchy.* A tree structure where each node of the tree represents a concept at some level of abstraction. Nodes toward the top of the tree are the most general. Leaf nodes represent individual data instances.

- *Conceptual clustering.* An incremental unsupervised clustering method that creates a concept hierarchy from a set of input instances.

- *Dendrogram.* A tree structure often used to depict a hierarchical clustering.

- *Dissimilarity matrix.* A matrix that contains values representing how different individual instances or clusters are from one another.

- *Incremental learning.* A form of learning that is supported in an unsupervised environment where instances are presented sequentially. As each new instance is seen, the learning model is modified to reflect the addition of the new instance.

- *Mixture.* A set of n probability distributions where each distribution represents a cluster.

- *Single linkage.* A method used with hierarchical clustering that defines distance as the shortest distance between a point in one cluster and a point in the second cluster.

EXERCISES

Review Questions

1. Compare and contrast conceptual and agglomerative clustering. Make a list of similarities and differences between the two approaches.

2. Compare and contrast the EM algorithm with the K-Means approach. List the similarities and differences between the two approaches.

3. Sum of squares values are often used for cluster evaluation. Between sum of squares was defined in Section 11.5 as the difference between total sum of squares and within sum of squares (*betweenss = totalss – withinss*). Explain why values of *betweenss/ totalss* closer to 1.0 offer positive support that the formed clusters characterize meaningful structure within a dataset.

Experimenting with R

1. Add statements to Script 11.1 to verify that the *betweenss/totss* value is 0.81.

2. Revisit Script 11.1 but use *fl* and *hr32* to cluster the data. Modify *xlim* and *ylim* as necessary in order for the clusters to properly display. Add statements to compute the value of *betweenss/totss*. Write a short summary of your results.

3. Experiment with Script 11.2 by varying the seed in an attempt to achieve a better clusters to classes result.

4. Use *hclust* to cluster the *Diabetes* dataset. Assume two classes in the data and perform a clusters to classes analysis. Compare the resultant confusion matrix with the confusion matrix seen in Script 11.2.

5. Apply RWeka's *GainRatioAttributeEval* function to the input attributes of the Diabetes dataset used in Script 11.2. Next, run Script 11.2 using all but the least useful input attribute. Repeat this process, each time eliminating the least effective remaining attribute. Write a short report that includes the confusion matrix for each experiment. Make a statement addressing the value of the input attributes in determining whether someone has diabetes.

6. Consider Script 11.3.

   a. Modify the script by calling *hclust* with method = average.

   b. Modify the script by calling *hclust* with method = single.

   c. Compare the dendrograms created in *a* and *b* with Figure 11.6.

7. Modify Script 11.5 by using *kmeans* and cardiologyNumerical to cluster the data. Summarize your results.

8. The RWeka package offers several clustering packages including SimpleKMeans and Cobweb. These packages are of interest as they allow you to use datasets containing categorical and mixed data types and automate data normalization. Use RWeka's SimpleKmeans to perform the K-Means experiments given in Script 11.1 and Script 11.2. A call to SimpleKMeans is of the form:

   SimpleKMeans(datafile, control=Weka_control(......))

   Use WOW(SimpleKMeans) to see all control options. Be sure to generate three clusters for the gamma-ray burst dataset. Output within-cluster mean and standard deviation values. Run your experiments several times varying the seed in order to obtain differing results. Summarize your results.

9. Use *kmeans* to cluster the sensor data (SensorData.csv). Use a decision tree or some other supervised technique to summarize your results.

10. Perform an unsupervised clustering with *hclust* and a dataset of your choice. Summarize your results.

11. Open the R file for Script 11.6 in RStudio. The script contains six values for the variable *card.data* where all but one value is commented. Each value extracts a subset of attributes from the original data. The subsets correspond to attribute relevance. For example, 8–11 and 15 represent the five best individual attributes as determined by *GainRatioAttributeEval*. Execute the script five times, each time using a different definition of *card.data*. Recording each resultant confusion matrix. Write a short report of your findings. List the attributes to use for building a best supervised model.

Computational Questions

1. Create tables to complete the agglomerative clustering example described in Section 11.2. Choose one of the techniques presented in Section 11.2 to pick a best clustering. As an alternative, develop your own method to make the choice. Explain why your choice represents a best clustering.

2. As part of our discussion of Script 11.4, we described a simple heuristic measure of attribute relevance computed by dividing the absolute value of the difference between attribute means by the global standard deviation for the attribute. Use this heuristic together with the mean values shown in Script 11.4 and the global hr32 standard deviation value to compute the attribute relevance scores for hr32.

3. Perform the third iteration of the K-Means algorithm for the example given in Section 11.1. What are the new cluster centers?

4. Consider the new instance I_8 with the attribute values listed below to be added to the concept hierarchy in Figure 11.3.

   Tails = Two
   Color = Dark
   Nuclei = Two

   a. Show the updated probability values for the root node once the instance has been entered into the hierarchy.

   b. Add the new instance to node n_2 and show the updated probability values for all affected nodes.

   c. Rather than having the instance become part of node n_2, assume the instance creates a new first-level node. Add the new node to the hierarchy and show the updated probability values for all affected nodes.

Installed Packages and Functions

Package Name	Function(s)
base / stats	*abs, aggregate, as.factor, barplot, cbind,cutTree, data.frame, dist, for, hclust, head ,if, kmeans, library, list, nrow, plot, points, round, sapply, scale, set.seed, sort, str, table*
Cluster	*daisy*
NbClust	*NbClust*
rpart	*rpart, rpart.plot, SimpleKMeans*
RWeka	*GainRatioAttributeEval*

## CHAPTER 12

A Case Study in Predicting Treatment Outcome

I N THIS CHAPTER, WE present a case study using actual patient data collected from "Ouch's back and fracture clinic" (OFC) introduced in Chapter 5. The original dataset contains records of 1330 patients (737 female) who underwent aggressive physical therapy between 2003 and 2008 for the treatment of low back injuries. The clinic name is fictitious, and patient identification information has been removed from the dataset.

As there are large differences in the strength, weight, and height levels of male and female patients, we decided to limit our experiments to the group of male patients, thereby making an analysis of the female patients a separate project. The female patient data is included in your supplementary materials.

Approximately 80% of the adult population suffers from acute low back syndrome (LBS) at some point during their life. Individuals with acute low back pain may seek relief from one or several sources including physical therapists, chiropractors, massage therapists, or primary care physicians. However, most cases of LBS self-correct within a period of 1–6 weeks without specific treatment.

Unfortunately, a small percentage of patients experience repeated occurrences of LBS where each incident is oftentimes slightly worse than the previous episode. Recurrence rates for these patients have been reported as high as 90% (Delitto et al., 1995). If physical therapy or chiropractic care is not helpful, patients may turn to acupuncture, injection, or medication. Regrettably, most of these treatments offer only temporary relief. Surgery designed to correct LBS can be dangerous and is not guaranteed to leave patients permanently pain free.

As an alternative to traditional therapies, OFC treatment methods are based on the theory that strength and pain are inversely related. Therefore, the purpose of an individual treatment program is to increase patient functional ability (strength, range of motion, and endurance) by totally isolating and aggressively strengthening the lumbar spine.

Exercise machines located in professional gyms purported to strengthen the lower back have now been found to do little more than work the hamstring and hip muscles. This was realized over 30 years ago when Arthur Jones made the first attempts to objectively measure knee and low back strength. Arthur Jones founded the MedX Corporation in 1972 with the sole purpose of designing machines to test and strengthen the muscles of the knee and low back. His theory claims a measureable inverse relationship between pain and strength as well as pain and flexibility. Hundreds of clinics worldwide now use the MedX machines first designed by Jones.

Figure 12.1 displays the core version of the larger lumbar extension MedX machine. The lumbar extension machine strengthens the lower lumbar spine by incorporating the pelvic restraint system shown in Chapter 5 (Figure 5.4). The MedX torso-rotation machine displayed in Figure 12.2 is also used by some patients to increase torso strength and flexibility.

The highly aggressive nature of the OFC treatment methodology has not been widely accepted by all physical therapists and medical doctors. Indeed, over 85% of patients first seek treatment via one or more of the aforementioned methods. Also, until recently, Medicare did not cover the treatment costs associated with this therapy. Their argument stated that aggressive therapy has not been proved to be effective for treating LBS.

In the next sections, we apply the seven-step knowledge discovery in data (KDD) process model outlined in Chapter 4 to a subset of this dataset consisting of 593 male patients. The KDD process includes steps for (1) goal identification, (2) target data creation, (3) data preprocessing, (4) data transformation, (5) data mining, (6) evaluation and interpretation, and (7) taking action.

FIGURE 12.1   The MedX Core lumbar extension machine.

FIGURE 12.2   The MedX torso-rotation machine.

12.1 GOAL IDENTIFICATION

After several meetings with the physical therapy staff, we developed two goals. Specifically,

- Identify patients early in treatment who are not likely to successfully complete their treatment program.

- Determine if significant differences in strength, flexibility, or pain levels exist between patients whose treatment program includes both lumbar extension and torso rotation and those patients whose program does not include torso rotation.

The rationale for the first goal is obvious. A certain percentage of individuals that begin treatment do not successfully complete their program. Patients may fail treatment for several reasons. A patient may drop out of the program if they feel they are not progressing in a positive way or if they are experiencing an increase in pain. Another possibility is that the patient's doctor decides that their treatment should be terminated. Finally, the patient may still experience significant pain at the completion of their treatment program.

The ability to identify an individual likely to fail in the first stages of treatment allows the therapist to make modifications to their program so as to give them a better chance at positive progress toward improvement. Also, if aggressive strength training is not a good alternative for a patient, it is best to know this early in the treatment program.

All patients treated for LBS use the lumbar extension machine as part of their treatment plan. However, only 35% of these same patients torso rotate. Whether or not a patient torso rotates is a decision made by their doctor or physical therapist. The overall health of the patient, the nature of the injury, and the philosophy of the medical staff all play a role as to whether a patient torso rotates. This fact motivated us to examine the differences between patients that do and do not torso rotate.

12.2 A MEASURE OF TREATMENT SUCCESS

When a patient is referred for treatment, she/he is examined by the physical therapist, a treatment plan specific to the patient is designed, and one or both of the MedX machines are used to test the patient's initial flexibility and strength levels. Also, at this initial visit, the patient self-reports their level of pain by filling out the Oswestry Low Back Disability Index (LBDI).

LBDI has been used for more than 35 years to measure disability levels in patients with LBS. LBDI is also employed to measure the degree of success relative to treatment outcome for LBS patients and is considered the "gold standard" of low back functional outcome tools (Fairbank and Pynsent, 2000).

An Oswestry LBDI score is determined by having the patient fill out a questionnaire containing ten categories. The patient selects one answer from a total of six possibilities (0 = no significant pain, 5 = severe pain) in each category. The LBDI categories include:

- Pain Intensity Now   Personal Care   Lifting
- Walking   Sitting   Standing
- Sleeping   Social Life   Traveling
- Changing Degree of Pain

To compute an individual patient's Oswestry score, the reported scores in all ten categories are summed and the total is multiplied by two. This allows for a minimum score of 0 and a maximum score of 100. The interpretation of a single LBDI score is accomplished by mapping a patient score to one of five range groups. Table 12.1 shows the range of Oswestry values for each of the five groups.

Minimal disability implies that the patient is able to cope with daily life without any special considerations other than exercising to maintain flexibility and strength. Patients experiencing moderate disability do experience some pain during sitting, lifting, and standing. These patients may not be able to travel long distances and may be unable to work. Pain is most often managed conservatively. Patients with long-term severe disability experience pain that severely affects their lifestyle. Patients in this group require detailed scrutiny.

TABLE 12.1   Oswestry Disability Index Measures

Score	Interpretation
0–20	Minimal disability
21–40	Moderate disability
41–60	Severe disability
61–80	Crippled
81–100	Bed-bound

Patients classified as crippled must seek positive intervention as back pain affects all aspects of their life. Patients don't often fall into the final category, but if they do, they are either confined to a bed or have exaggerated their symptoms. As the LBDI is a self-report of a patient's perceived disability, an LBDI score has a degree of subjectivity. The range of LBDI scores for the current research is 0–76. For our case study, we defined treatment success as the patient reporting a final Oswestry score of 20 or less.

12.3 TARGET DATA CREATION

Data was gathered from the records of patients who were treated during the years 2003–2008. Over 75% of patient records were not initially stored electronically. Several part-time individuals were involved in transferring records to an electronic form. Each LBS patient record is represented by 90 attributes. A summarized list of these attributes is shown in Table 12.2.

Diagnosis, previous treatments, and previous health history represent categories containing 35, 6, and 23 attributes, respectively.

The lumbar extension and trunk rotation attributes in Table 12.2 provide patient strength (lumbar extension weight and trunk rotation weight) and flexibility (lumbar extension rom and trunk rotation rom) values as measured by the MedX machines. The diagnosis category displayed in the table actually represents a collection of 35 attributes. All diagnosis attributes are categorical (yes, no) and tell us whether an individual patient is or is not

TABLE 12.2   Clinical Attributes

Gender (*male, female*)
Patient age
Patient height
Patient weight
Referral (*MD, Self, DC, workers comp., other*)
Onset (*acute, sub-acute, chronic*)
Diagnosis (*35 attributes*)
Disability (*yes, no*)
Previous treatments (*six attributes*)
Employment status (*six values*)
Time unemployed (*six values*)
Smoker (*yes, no*)
Previous health history (*23 attributes*)
Insurance type (*five values*)
Number of visits
Lumbar extension ROM start
Lumbar extension ROM end
Lumbar extension weight start
Lumbar extension weight end
Trunk rotation ROM start
Trunk rotation ROM end
Trunk rotation weight start
Trunk rotation weight end
Treatment ended (*six values*)
Patient satisfaction (*five values*)
Oswestry LBDI start
Oswestry LBDI end

afflicted with a specific diagnostic condition. The great majority of patients do not suffer from any of these conditions. Finally, the previous health history category tells us whether the patient had or currently suffers from one or more of 23 possible diseases including arthritis, neurological conditions, stroke, heart disease, and the like. Most patients do not suffer from any of these conditions.

12.4 DATA PREPROCESSING

As we used Oswestry LBDI for our measure of treatment outcome, our next step was to eliminate data instances not having a value for Oswestry start and Oswestry end. Of the 593 male patients, 231 showed a measure for Oswestry start and only 172 had a value for both measures. Numerous reasons were cited for this. The most common being inconsistencies in requiring patients to fill out the Oswestry questionnaire, patients that decided against starting treatment, patient transfers, and patients ending treatment without notification. Also, five of the remaining patients did not show values for lumbar extension in and out. These patients were also eliminated from the data, thereby leaving 226 instances with a value for Oswestry start and 167 instances showing a value for both measures.

12.5 DATA TRANSFORMATION

The initial data was stored numerically in an Excel spreadsheet. As more than 75 of the attributes represent categorical data, we converted all but 14 numeric representations to their actual categorical value.

We deleted all *yes-no* input attributes showing less than two occurrences of either yes or no. Next, we eliminated four trunk rotation attributes as less than 60 of the 167 instances showed trunk rotation values. Attribute correlations were negligible with the exception of the 0.52 correlation between Oswestry end and Oswestry start.

As our first goal was predictive, we limited our data to those attributes whose values were known at the start of treatment. This eliminated lumbar extension end and lumbar extension range of motion (ROM) end, patient satisfaction, and number of visits. The remaining 46 input attributes were used to build the models for predicting Oswestry end.

12.6 DATA MINING

Two thirds of the data was randomly selected for training with the remaining third used for model testing. Our initial experiments using a variety of machine learning techniques with both the numeric and mixed forms of the dataset were unable to accurately determine Oswestry end. This motivated us to apply a data transformation whereby Oswestry end values less than or equal to 20 were replaced with 0 and values greater than 20 where given a value of 1. This transformation corresponds with our definition of treatment success.

12.6.1 Two-Class Experiments

To establish a rough benchmark for the two-class problem, we applied the *knn nearest neighbor* function located in the *class* package to a dataset consisting of the original numeric input attributes whose values are known prior to treatment. The column of Oswestry end values were replaced with a corresponding column of zeros and ones.

The nearest neighbor approach stores instances rather than a generalized model of the data. A new instance to be classified is compared to each stored instance. The instance is given the classification of the stored instance(s) of greatest similarity. The algorithm requires numeric input data, a categorical output value, and by default employs simple Euclidean distance to measure similarity.

The format used by *knn* is not standard and is best illustrated by example. Here is the call and resultant output for our benchmark experiment.

```
>Library(class)
>my.knn <-knn(Of.train[,-1],Of.test[,-1],Of.train[,1],k=3)
> y<-table(my.knn,Of.test[,1])
> y

my.knn  0   1
     0 31  18
     1  4   3

> confusionP(y)
  Correct= 34
Incorrect= 22
Accuracy = 60.71 %
```

The first argument to *knn* is the training data with the output attribute (column 1) removed. The second argument is the test data minus the output attribute. The third argument is the output attribute as represented in the training data. The final argument tells *knn* how many nearest neighbors to use for the classification. With $k = 3$, a new instance is classified with a majority of three rule.

The *table* function displays the confusion matrix with columns and rows interchanged. That is, 21 total test set instances are from the failed outcome group. Of the 21, 18 were incorrectly classified as having a successful outcome. This less than optimal result is to be expected as the data is not in its preprocessed form and contains many irrelevant attributes. Let's see if we can do better than the 60% benchmark and in particular improve on the 18 incorrectly classified failures.

For the two-class experiments with the preprocessed data, we used *JRip(RWeka)*, *MLP(RWeka)*, and *randomForest(randomForest)*. The test set results for all three methods were similar with the *randomForest* showing a best overall test set accuracy of 78.57%. Here is the confusion matrix seen for the *randomForest* experiment:

```
              Predicted
    actual    0   1
          0  38   5
          1   7   6
```

```
> confusionP(credit.perf)
  Correct= 44 Incorrect= 12
  Accuracy = 78.57 %
```

The JRip covering rule algorithm generated one rule with an antecedent condition and a default rule as given below.

```
(oswest.in >= 42) => success=1 (29.0/8.0)
            => success=0 (82.0/14.0)
```

Recall the regression tree of Chapter 6 created with the 48 instance subset of these data made the first tree split on Oswestry start (oswest.in) = 42.

As a final experiment, we included the instances with an ostwestry start but no Oswestry end. We gave these instances a 1 for Oswestry end indicating failure. Once again, the random forest showed the best result with a 77.63% test set accuracy. More importantly, 24 of the 30 (80%) test set instance failures were correctly identified.

12.7 INTERPRETATION AND EVALUATION

The computed mean Oswestry score for patients completing the treatment program decreased from 31.57 to 16.09. However, our initial experiments attempting to predict treatment outcome as measured by raw Oswestry end scores were unsuccessful. A variety of possibilities exist including a lack of relevant input attributes. As a second attempt to predict failures, we replaced the Oswestry output attribute with a new binary output attribute with value 0 for successful patients and 1 for patients with final oswestry values greater than 20. We also included instances not showing a value for Oswestry end as failures. As a result, we were able to identify 80% of those test set patients who did not achieve a positive result.

12.7.1 Should Patients Torso Rotate?

All patients treated for LBS use the lumbar extension machine as part of their treatment plan. However, less than 40% of these patients torso rotate. Whether or not a patient torso rotates is a decision made by their doctor or physical therapist. The overall health of the patient, the nature of the injury, and the philosophy of the medical staff all play a role as to whether the patient torso rotates. This fact motivated us to examine the differences between patients that do and do not torso rotate.

Table 12.3 compares the lumbar extension start and end weight values for male LBS patients that did and did not torso rotate. For each group, within-group paired comparison t tests of significance (start vs. end weight) were significant ($p \leq 0.01$). In addition, the difference between mean end values of the two groups was significant ($p \leq 0.01$). That is, torso rotating patients were able to lumbar extend with significantly more weight when compared to the group that did not rotate (143.76 vs. 125.78). This observation must be taken with caution as the group that torso rotated had significantly higher initial weight values.

Table 12.3 also displays the percent increase in measured strength for each group. Notice that the patients who did not torso rotate increased their lumbar extension weight by more than 95% whereas the torso rotating group showed an increase of 56%. This comparison

TABLE 12.3 Lumbar Extension Weight Values for Male Patients Who Did and Did Not Torso Rotate

	Torso Rotation Patients (191 Total)	Patients Did Not Torso Rotate (304 Total)
Start value	Mean: 99.24	Mean: 68.34
	SD: 31.76	SD: 24.70
End value	Mean: 143.76	Mean: 125.78
	SD: 33.67	SD: 37.84
Percent increase (paired comparison start vs. end)	Mean: 56.16%	Mean: 95.99%
	Median: 38.64%	Median: 89.23%

must also be made with caution as the placement of patients within each group was not random. An ideal experiment would apply a random selection of patients to each group. In this way, the comparison between the two groups allows for cause and effect conclusions.

12.8 TAKING ACTION

This case study examined the treatment outcome of male patients who underwent aggressive LBS therapy. Aggressive treatment of LBS is based on the theory that lower back pain is inversely related to spinal strength and flexibility. Our results clearly support the success of aggressive therapy for LBS.

A primary goal of this research was to build a model able to identify those patients likely to perform poorly in their treatment program. Once identified, appropriate modifications to individual treatment programs can be made in hopes of a better treatment outcome. Although we experienced some success in this endeavor, we did not meet our expectations of building a highly accurate predictive model. Our lack of data, crisp definition of treatment success as well as the subjective nature of the Oswestry scoring mechanism may all be partially to blame. This fact motivates us to focus on collecting and building models using patient data taken at the start of treatment and again after 2 or 3 weeks of treatment. In doing so, we may see added success in identifying patients whose current treatment plan is not leading to a best result.

12.9 CHAPTER SUMMARY

Our purpose here was to give you a basic idea about how you might go about an analytics process with your own data. Your supplementary materials include *Casedata.zip* which contains four Excel files to use for further experimentation with this data. Namely,

- Definitions: Lists all attributes and possible values for the original data.

- FemaleNumericComplete: Contains all 757 female instances in numeric format.

- MaleNumericComplete: Contains all 593 male instances in numeric format.

- MalePreprocessed: Contains the preprocessed data used for the case study.

You are encouraged to use these files to repeat the experiments given above, form and test new hypotheses, and develop your report writing skills!

Bibliography

Agrawal, R., Imielinski, T., and Swami, A. (1993). Mining Association rules between sets of items in large databases. In P. Buneman and S. Jajordia, eds., *Proceedings of the ACM Sigmoid International Conference on Management of Data*. New York: ACM.

Baltazar, H. (1997). Tracking telephone fraud fast. *Computerworld*, 31, 11, 75.

Baltazar, H. (2000). NBA coaches' latest weapon: Data mining. *PC Week*, March 6, 69.

Baumer, B.S., Kaplan, D. T., and Horton, N.J. (2017). *Modern Data Science with R*. Boca Raton, FL: CRC Press.

Blair, E., and Tibshirani, R. (2003). Machine learning methods applied to DNA microarray data can improve the diagnosis of cancer. *SIGKDD Explorations*, 5, 2, 48–55.

Boser, B. E., et al. (1992). A training algorithm for optimal margin classifiers. *Proceedings of the Fifth Annual Workshop on Computational Learning Theory*, 5, 144–152.

Brachman, R. J., Khabaza, T., Kloesgen, W., Pieatetsky-Shapiro, G., and Simoudis, E. (1996). Mining business databases. *Communications of the ACM*, 39, 11, 42–48.

Breiman, L. (1996). Bagging predictors. *Machine Learning*, 24, 2, 123–140.

Breiman, L. (2001). Random forests, *Machine Learning*, 45, 1, 5–32.

Breiman, L., Friedman, J., Olshen, R., and Stone, C. (1984). *Classification and Regression Trees*. Monterey, CA: Wadsworth International Group.

Burges, C. (1998). A tutorial on support vector machines for pattern recognition. *Data Mining and Knowledge Discovery* 2, 121–167.

Calders, T., Dexters N., and Goethals, B. (2007). Mining frequent itemsets in a stream. In *Proceedings of the 2007 Seventh IEEE International Conference on Data Mining*, ICDM'07, Washington, DC. IEEE Computer Society, 83–92.

Case, S., Azarmi, N., Thint, M., and Ohtani, T. (2001). Enhancing E-communities with agent-based systems. *Computer*, July, 64–69.

Cendrowska, J. (1987). PRISM: An algorithm for inducing modular rules. *International Journal of Man-Machine Studies*, 27, 4, 349–370.

Chester, M. (1993). *Neural Networks—A Tutorial*. Upper Saddle River, NJ: PTR Prentice Hall.

Chou, W. Y. S., Hunt, Y. M., Beckjord, E. B., Moser, R. P., and Hesse, B. W. (2009). Social media use in the United States: Implications for health communication. *Journal of Medical Internet Research*, 11, 4, e48.

Civco, D. L. (1991). Landsat TM Land use and land cover mapping using an artificial neural network. In *Proceedings of the 1991 Annual Meeting of the American Society for Photogrammetry and Remote Sensing*, Baltimore, MD, 3, 66–77.

Cohen, W., (1995). Fast effective rule induction. In *12th International Conference on Machine Learning*, Washington D.C., USA, 115–123.

Cox, E. (2000). Free-Form text data mining integrating fuzzy systems, self-organizing neural nets and rule-based knowledge bases. *PC AI*, September–October, 22–26.

Culotta, A. (2010). Towards detecting influenza epidemics by analyzing Twitter messages. In *Proceedings of the First Workshop on Social Media Analytics*, July. ACM, 115–122.

Dasgupta, A., and Raftery, A. E. (1998). Detecting features in spatial point processes with clutter via model-based clustering. *Journal of the American Statistical Association*, 93, 441, 294–302.

Dawson, R. (1995). The "unusual episode" data revisited. *Journal of Statistics Education*, 3, 3.

Delitto, A., Erhard, R.E., and Bowling R.W. (1995). A treatment-based classification approach to low back syndrome, identifying and staging patients for conservative treatment. *Physical Therapy*, 75, 6, 470–489.

Dempster, A. P., Laird, N. M., and Rubin, D. B. (1977). Maximum-likelihood from incomplete data via the EM algorithm (with Discussion). *Journal of the Royal Statistical Society, Series B*, 39, 1, 1–38.

Dixon, W. J. (1983). *Introduction to Statistical Analysis*, 4th ed. New York: McGraw-Hill.

Domingos, P., and Hulten, G. (2000). Mining high speed data streams. In *Proceedings of the 6th ACM SIGKDD International Conference on Knowledge Discovery and Data Mining (KDD'00)*, New York. ACM, 71–80.

Duda, R., Gaschnig, J., and Hart, P. (1979). Model design in the PROSPECTOR consultant system for mineral exploration. In D. Michie, ed., *Expert Systems in the Microelectronic Age*. Edinburgh, Scotland: Edinburgh University Press, 153–167.

Durfee, E. H. (2001). Scaling up agent coordination strategies. *Computer*, July, 39–46.

Dwinnell, W. (1999). Text mining dealing with unstructured data. *PC AI*, May–June, 20–23.

Edelstein, H. A. (2001). Pan for gold in the clickstream. *Information Week*, March 12.

Fairbanks, J.C., and Pynsent, P.B. (2000). The Oswestry disability index. *Spine*, 25, 22, 2940–2952.

Fayyad, U., Haussler, D. and Stolorz, P. (1996). Mining scientific data. *Communications of the ACM*, 39, 11, 51–57.

Fisher, D. (1987). Knowledge acquisition via incremental conceptual clustering. *Machine Learning*, 2, 2, 139–172.

Frank, E., Hall, M.A., and Witten, I. A. (2016). *The WEKA Workbench. Online Appendix for Data Mining: Practical Machine Learning Tools and Techniques*, 4th ed. San Francisco, CA: Morgan Kaufmann.

Freund, Y., and Schapire, R. E. (1996). Experiments with a new boosting algorithm. In L. Saitta, ed., *Proc. Thirteenth International Conference on Machine Learning*. San Francisco, CA: Morgan Kaufmann, 148–156.

Ganti, V., Gehrke, J., and Ramakrishnan, R. (1999). Mining very large database. *Computer*, August, 38–45.

Gardner, S. R. (1998). Building the data warehouse. *Communications of the ACM*, 41, 9, 52–60.

Gehrke, J., Ramakrishnan, R., and Ganti, V. (2000). RainForest—a framework for fast decision tree construction of large datasets. *Data Mining and Knowledge Discovery* 4, 127–162.

Gennari, J. H., Langley, P., and Fisher, D. (1989). Models of incremental concept formation. *Artificial Intelligence*, 40, (1–3), 11–61.

Giarratano, J., and Riley, G. (1989). *Expert Systems: Principles and Programming*. New York: PWS-Kent.

Gill, H. S., and Rao, P. C. (1996). *The Official Guide to Data Warehousing*. Indianapolis, IN: Que Publishing.

Giovinazzo, W. A. (2000). *Object-Oriented Data Warehouse Design (Building a Star Schema)*. Upper Saddle River, NJ: Prentice Hall.

Granstein, L. (1999). Looking for patterns. *Wall Street Journal*, June 21.

Grossman, R.L., Hornick, M.F., Meyer, G. (2002). Data mining standards initiatives, *Communications of the ACM*, 45, 8, 59–61.

Haag, S., Cummings, M., and McCubbrey, D. (2002). *Management Information Systems for the Information Age*, 3rd ed. Boston, MA: McGraw-Hill.

Haglin, D. J. and Roiger. R. J. (2005). A tool for public analysis of scientific data, *Data Science Journal*, 4, 39–52.

Hornik, K., Buchta, C., and Zeileis, A. (2009). Open-source machine learning: R meets Weka. *Computational Statistics*, 24, 2, 225–232, doi;10.1007/s00180-008-0119-7

Hosmer, D. W., and Lemeshow, S. (1989). *Applied Logistic Regression*. New York: John Wiley & Sons.

Huntsberger, D. V. (1967). *Elements of Statistical Inference*. Boston, MA: Allyn and Bacon.

Inmon, W. (1996). *Building the Data Warehouse*. New York: John Wiley & Sons.

Jain, A. K., Mao, J., and Mohiuddin, K. M. (1996). Artificial neural networks: A tutorial. *Computer*, March, 31–44.

Jones, A. (1993). My first half century in the iron game. *Ironman Magazine*, July, 94–98.

Kabacoff, R. I. (2015). *R In Action: Data Analysis and graphics with R*, 2rd ed. New York: Manning Publications.

Kass, G. V. (1980). An exploratory technique for investigating large quantities of categorical data. *Applied Statistics*, 29, IS, 119–127.

Kaur, I., Mann, D. (2014). Data mining in cloud computing. *International Journal of Advanced Research in Computer Science and Software Engineering*, 4, 3, 1178–1183.

Khan, J. et al. (2001). Classification and diagnostic prediction of cancers using gene expression profiling and artificial neural networks. *Nature Medicine*, 7, 673–679.

Kimball, R., Reeves, L., Ross, M., and Thornthwaite, W. (1998). *The Data Warehouse Lifecycle Toolkit: Expert Methods for Designing, Developing, and Deploying Data Warehouses*. New York: John Wiley & Sons.

Kohonen, T. (1982). Clustering, taxonomy, and topological maps of patterns. In M. Lang, ed., *Proceedings of the Sixth International Conference on Pattern Recognition*. Silver Spring, MD: IEEE Computer Society Press, 114–125.

Kudyba, S. (2014). *Big Data, Mining and Analytics*. Boca Raton, FL: CRC Press.

Larose, D.T., and Larose, C.D. (2015). *Data Mining and Predictive analytics*, 2nd ed. New York: John Wiley & Sons.

Lashkari, Y., Metral, M., and Maes, P. (1994). Collaborative interface agents. In *Proceedings of the Twelfth National Conference on Artificial Intelligence*. Menlo Park, CA: American Association of Artificial Intelligence, 444–450.

Lin, N. (2015). *Applied Business Analytics*. Upper Saddle River, NJ: Pearson.

Lloyd, S. P. (1982). Least squares quantization in PCM. *IEEE Transactions on Information Theory*, 28, 2, 129–137.

Long, S. L. (1989). *Regression Models for Categorical and Limited Dependent Variables*. Thousand Oaks, CA: Sage Publications Inc.

Maclin, R., and Opitz, D. (1997). An empirical evaluation of bagging and boosting. In *Fourteenth National Conference on Artificial Intelligence*. Providence, RI: AAAI Press.

Maiers, J., and Sherif, Y. S. (1985). Application of fuzzy set theory. *IEEE Transactions on Systems, Man, and Cybernetics, SMC*, 15, 1, 41–48.

Manganaris, S. (2000). Estimating intrinsic customer value. *DB2 Magazine*, 5, 3, 44–50.

Manning, A. (2015). *Databases for Small Business*. New York: Springer.

McCulloch, W. S., and Pitts, W. (1943). A logical calculus of the ideas imminent in nervous activity. *Bulletin of Mathematical Biophysics*, 5, 115–137.

Mena, J. (2000). Bringing them back. *Intelligent Enterprise*, 3, 11, 39–42.

Merril, D. M., and Tennyson, R. D. (1977). *Teaching Concepts: An Instructional Design Guide*. Englewood Cliffs, NJ: Educational Technology Publications.

Mitchell, T. M. (1997). Does machine learning really work? *AI Magazine*, 18, 3, 11–20.

Mobasher, B., Cooley, R., and Srivastava, J. (2000). Automatic personalization based on web usage mining. *Communications of the ACM*, 43, 8, 142–151.

Mone, G. (2013) Beyond Hadoop. *Communications of the ACM*, 56, 1, 22–24.

Mukherjee, S., Feigelson, E. D., Babu, G. J., Murtagh, F., Fraley, C., and Rafter, A. (1998). Three types of gamma ray bursts. *Astrophysical Journal*, 508, 1, 314–327.

Ortigosa, A., Carro, R. M., and Quiroga, J. I. (2014). Predicting user personality by mining social interactions in Facebook. *Journal of Computer and System Sciences*, 80, 1, 57–71.

Peixoto, J. L. (1990). A property of well-formulated polynomial regression models. *American Statistician*, 44, 26–30.

Perkowitz, M., and Etzioni, O. (2000). Adaptive web sites. *Communications of the ACM*, 43, 8, 152–158.

Piegorsch, W. W. (2015). *Statistical Data Analytics*. New York: John Wiley & Sons.

Platt, J.C. (1998). Fast training of support vector machines using sequential minimal optimization. In B. Schoelkopf, C. Burgers, and A. Smola, eds., *Advances in Kernel Methods – Support Vector Learning*. Cambridge, MA: MIT Press.

Quinlan, J. R. (1986). Induction of decision trees. *Machine Learning*, 1, 1, 81–106.

Quinlan, J. R. (1993). *Programs for Machine Learning*. San Mateo, CA: Morgan Kaufmann.

Quinlan, J. R. (1994). Comparing connectionist and symbolic learning methods. In S. J. Hanson, G. A. Drastall, and R. L. Rivest, eds., *Computational Learning Theory and Natural Learning Systems*. Cambridge, MA: MIT Press, 445–456.

Rich, E., and Knight, K. (1991). *Artificial Intelligence*, 2nd ed. New York: McGraw-Hill.

Roiger, R.J. (2005). Teaching an introductory course in data mining. In *Proceedings of the 10th Annual SIGCSE Conference on Innovation and Technology in Computer Science Education*, ACM Special Interest Group on Computer Science Education, Universidade Nova de Lisboa.

Roiger, R.J. (2016). *Data Mining a Tutorial-Based Primer*, 2nd ed. Boca Raton, FL: Chapman and Hall/CRC.

Rowley, J. (2007). The wisdom hierarchy: Representations of the DIKW hierarchy. *Journal of Information and Communication Science*, 33, 2, 163–180.

Salkind, N. J. (2012). *Exploring Research*, 8th ed. Boston, MA: Pearson.

Schmidt, C. W. (2012). Trending now: Using social media to predict and track disease outbreaks. *Environmental Health Perspectives*, 120, 1, a30.

Senator, T. E., Goldbert, H. G., Wooten, J., Cottini, M. A., Khan, A. F. U., Klinger, C. D., Llamas, W. M., Marrone, M. P., and Wong, R. W. H. (1995). The financial crimes enforcement network AI system (FAIS): Identifying potential money laundering from reports of large cash transactions. *AI Magazine*, 16, 4, 21–39.

Shafer, J., Agrawal, R., and Mehta, M. (1996). SPRINT: A scalable parallel classifier for data mining. In *Proceedings of the 22nd VLDB Conference Mumbai(Bombay), India*, 544–555.

Shannon, C. E. (1950). Programming a computer for playing chess. *Philosophical Magazine*, 41, 4, 256–275.

Shavlik, J., Mooney, J., and Towell, G. (1990). Symbolic and neural learning algorithms: An experimental comparison (Revised). Tech. Rept. No. 955, Computer Sciences Department, University of Wisconsin, Madison, WI.

Shortliffe, E. H. (1976). *MYCIN: Computer-Based Medical Consultations*. New York: Elsevier Press.

Signorini, A., Segre, A. M., and Polgreen, P. M. (2011). The use of Twitter to track levels of disease activity and public concern in the US during the influenza A H1N1 pandemic. *PLoS One*, 6, 5, e19467.

Spiliopoulou, M. (2000). Web usage mining for web site evaluation. *Communications of the ACM*, 43, 8, 127–134.

Sycara, K. P. (1998). The many faces of agents. *AI Magazine*, 19, 2, 11–12.

Thuraisingham, B. (2003). *Web Data Mining and Applications in Business Intelligence and Counter-Terrorism*. Boca Raton, FL: CRC Press.

Torgo, L. (2011). *Data Mining with R: Learning with Case Studies*. Boca Raton, FL: CRC Press.

Turing, A. M. (1950). Computing machinery and intelligence. *Mind* 59, 433–460.

Vafaie, H., and DeJong, K. (1992). Genetic algorithms as a tool for feature selection in machine learning. In *Proc. International Conference on tools with Artificial Intelligence*, Arlington, VA. IEEE Computer Society Press, 200–205.

Vapnik, V. (1998). *Statistical Learning Theory*. New York: John Wiley & Sons.

Vapnik, V. (1999). *The Nature of Statistical Learning Theory*, 2nd ed. New York: Springer.

Weiss, S. M., and Indurkhya, N. (1998). *Predictive Data Mining: A Practical Guide*. San Francisco, CA: Morgan Kaufmann.

Widrow, B., and Lehr, M. A. (1995). Perceptrons, adalines, and backpropagation. In M.A. Arbib, ed., *The Handbook of Brain Theory and Neural Networks*. Cambridge, MA: MIT Press, 719–724.

Widrow, B., Rumelhart, D. E., and Lehr, M. A. (1994). Neural networks: Applications in industry, business and science. *Communications of the ACM*, 37, 3, 93–105.

Wilson, C., Boe, B., Sala, A., Puttaswamy, K. P., and Zhao, B. Y. (2009, April). User interactions in social networks and their implications. In *Proceedings of the 4th ACM European Conference on Computer Systems*. ACM, Nuremberg, Germany 205–218.

Winston, P. H. (1992). *Artificial Intelligence*, 3rd ed. Reading, MA: Addison-Wesley.

Witten, I. H., Frank, E. and Hall, M. (2011). *Data Mining: Practical Machine Learning Tools and Techniques*, 3rd ed. San Francisco, CA: Morgan Kaufmann.

Wu, X., Kumar, V. (2009). *The Top 10 Algorithms in Data Mining*. Boca Raton, FL: Chapman and Hall/CRC.

Zadeh, L. (1965). Fuzzy sets. *Information and Control*, 8, 3, 338–353.

Zikopoulos, P.C., Eaton, C., deRoos, D., Deutsch, T., and Lapis, G. (2012). *Understanding Big Data-Analytics for Enterprise Class Hadoop and Streaming Data*. New York: McGraw Hill.

Appendix A

Supplementary Materials and More Datasets

SUPPLEMENTARY MATERIALS

The datasets, scripts, and functions written for your text are contained in a zip file available at two locations:

- The CRC website https://www.crcpress.com/9780367439149

- http://krypton.mnsu.edu/~sa7379bt/

If you have any problems obtaining or using these materials or if you have questions or concerns about the text, please email me at richard.roiger@mnsu.edu I will be more than happy to assist!

MORE DATASETS

KDNuggets

KDNuggets is the leading information repository for data mining and knowledge discovery. The site includes topics on data mining—companies, software, publications, courses, datasets, and more. Here is the home page of the KDNuggets Web site:

http://www.kdnuggets.com

Here is a link to popular datasets from several domains.

http://www.kdnuggets.com/datasets/index.html

Machine Learning Repository

The UCI Machine Learning Repository contains a wealth of data from several domains. Here is the UCI home page address.

https://archive.ics.uci.edu/ml/datasets.html

Here is the ftp archive site for downloading the datasets in the UCI library.

ftp://ftp.ics.uci.edu/pub/machine-learning-databases/

Community Datasets—IBM Watson Analytics

IBM Watson Analytics is a commercial cloud-based predictive analytics and data visualization tool. Several interesting Excel-based datasets for data mining are available through the Watson analytics community at

https://community.watsonanalytics.com/guide-to-sample-datasets/

These datasets can be used without modification. You can also opt for a free trial of Watson analytics by visiting https://community.watsonanalytics.com/.

Appendix B

Statistics for Performance Evaluation

SINGLE-VALUED SUMMARY STATISTICS

The *mean* or average value is computed by summing the data and dividing the sum by the number of data items. Specifically,

$$\mu = \frac{1}{n}\sum_{i=1}^{n} x_i \qquad (B.1)$$

where
μ is the mean value
n is the number of data items
X_i is the ith data item.

Here is a formula for computing variance.

$$\sigma^2 = \frac{1}{n}\sum_{i=1}^{n}(\mu - x_i)^2 \qquad (B.2)$$

where
σ^2 is the variance
μ is the population mean
n is the number of data items
X_i is the ith data item.

When calculating the variance for a sampling of data, a better result is obtained by dividing the sum of squares by $n - 1$ rather than by n. The proof of this is beyond the scope of this book; however, it suffices to say that when the division is by $n - 1$, the sample variance is an unbiased estimator of the population variance. An unbiased estimator has the characteristic that the average value of the estimator taken over all possible samples is equal to the parameter being estimated.

THE NORMAL DISTRIBUTION

Here is the equation for the normal or bell-shaped curve.

$$f(x) = 1/(\sqrt{2\pi}\sigma)e^{-(x-\mu)^2/2\sigma^2} \tag{B.3}$$

where

$f(x)$ is the height of the curve corresponding to values of x

e is the base of natural logarithms approximated by 2.718282

μ is the arithmetic mean for the data

σ is the standard deviation.

COMPARING SUPERVISED MODELS

In Chapter 9, we described a general technique for comparing two supervised learner models using the same test dataset. Here we provide two additional techniques for comparing supervised models. In both cases, model test set error rate is treated as a sample mean.

Comparing Models with Independent Test Data

With two independent test sets, we simply compute the variance for each model and apply the classical hypothesis testing procedure. Here's an outline of the technique.

Given,

- Two models M_1 and M_2 built with the same training data

- Two independent test sets, set A containing n_1 elements and set B with n_2 elements

- Error rate E_1 and variance v_1 for model M_1 on test set A

- Error rate E_2 and variance v_2 for model M_2 on test set B

Compute,

$$P = \frac{|E_1 - E_2|}{\sqrt{(v_1/n_1 + v_2/n_2)}} \tag{B.4}$$

Conclude,

- If $P \geq 2$, the difference in the test set performance of model M_1 and model M_2 is significant.

Let's look at an example. Suppose we wish to compare the test set performance of learner models, M_1 and M_2. We test M_1 on test set A and M_2 on test set B. Each test set contains

100 instances. M_1 achieves an 80% classification accuracy with set A and M_2 obtains a 70% accuracy with test set B. We wish to know if model M_1 has performed significantly better than model M_2. Here are the computations.

- For model M_1:

  $E_1 = 0.20$

  $v_1 = 0.2\,(1 - 0.2) = 0.16$

- For model M_2:

  $E_2 = 0.30$

  $v_2 = 0.3\,(1 - 0.3) = 0.21$

- The computation for P is:

$$P = \frac{|0.20 - 0.30|}{\sqrt{(0.16/100 + 0.21/100)}}$$

$$P \approx 1.4714$$

Since $P < 2$, the difference in model performance is not considered to be significant. We can increase our confidence in the result by switching the two test sets and repeating the experiment. This is especially important if a significant difference is seen with the initial test set selection. The average of the two values for P is then used for the significance test.

Pairwise Comparison with a Single Test Data Set

When the same test set is applied to the data, one option is to perform an instance-by-instance pairwise matching of the test set results. With an instance-based comparison, a single variance score based on pairwise differences is computed. Here is the formula for calculating the joint variance.

$$V_{12} = \frac{1}{n-1} \sum_{i=1}^{n} [(e_{1i} - e_{2i}) - (E_1 - E_2)]^2 \tag{B.5}$$

where
    V_{12} is the joint variance
    e_{1i} is the classifier error on the ith instance for learner model 1
    e_{2i} is the classifier error on the ith instance for learner model 2
    $E_1 - E_2$ is the overall classifier error rate for model 1 minus the classifier error rate for model 2
    n is the total number of test set instances.

When test set error rate is the measure by which two models are compared, the output attribute is categorical. Therefore, for any instance i contained in class j, e_{ij} is 0 if the classification is correct and 1 if the classification is in error. When the output attribute is numerical, e_{ij} represents the absolute difference between the computed and actual output value. With the revised formula for computing joint variance, the equation to test for a significant difference in model performance becomes

$$P = \frac{|E_1 - E_2|}{\sqrt{V_{12}/n}} \tag{B.6}$$

Once again, a 95% confidence level for a significant difference in model test set performance is seen if $P \geq 2$. The above technique is appropriate only if an instance-based pairwise comparison of model performance is possible. In the next section, we address the case where an instance-based comparison is not possible.

Confidence Intervals for Numeric Output

Just as when the output was categorical, we are interested in computing confidence intervals for one or more numeric measures. For purposes of illustration, we use *mean absolute error*. As with classifier error rate, *mean absolute error* is treated as a sample mean. The sample variance is given by the formula:

$$\text{variance}(mae) = \frac{1}{n-1} \sum_{i=1}^{n} (e_i \ mae)^2 \tag{B.7}$$

where
　e_i is the absolute error for the ith instance
　n is the number of instances.

Let's look at an example using the data in Table 9.2. To determine a confidence interval for the *mae* computed for the data in Table 9.2, we first calculate the variance. Specifically,

$$\text{variance}(0.0604) = \frac{1}{14} \sum_{i=1}^{15} (e_i - 0.0604)^2$$

$$\approx (0.024 - 0.0604)^2 + (0.002 - 0.0604)^2 + \cdots + (0.001 - 0.0604)^2$$

$$\approx 0.0092$$

Next, as with classifier error rate, we compute the standard error for the *mae* as the square root of the variance divided by the number of sample instances.

$$SE = \sqrt{(0.0092/15)} \approx 0.0248$$

Finally, to calculate the 95% confidence interval by, respectively, subtracting and adding two standard errors to the computed *mae*. This tells us we can be 95% confident that the actual *mae* falls somewhere between 0.0108 and 0.1100.

Comparing Models with Numeric Output

The procedure for comparing models giving numeric output is identical to that for models with categorical output. In the case where two independent test sets are available and *mae* measures model performance, the classical hypothesis testing model takes the form:

$$P = \frac{|mae_1 - mae_2|}{\sqrt{(v_1/n_1 + v_2/n_2)}} \tag{B.8}$$

where
    mae_1 is the mean absolute error for model M_1
    mae_2 is the mean absolute error for model M_2
    V_1 and V_2 are variance scores associated with M_1 and M_2
    n_1 and n_2 are the number of instances within each respective test set.

When the models are tested on the same data and a pairwise comparison is possible, we use the formula:

$$P = \frac{|mae_1 - mae_2|}{\sqrt{V_{12}/n}} \tag{B.9}$$

where
    mae_1 is the mean absolute error for model M_1
    mae_2 is the mean absolute error for model M_2
    V_{12} is the joint variance computed with the formula defined in B.5
    n is the number of test set instances.

When the same test data is applied but a pairwise comparison is not possible, the most straightforward approach is to compute the variance associated with the *mae* for each model using the equation:

$$\text{variance}(mae_j) = \frac{1}{n-1}\sum_{i=1}^{n}(e_i \ mae_j)^2 \tag{B.10}$$

where
    mae_j is the mean absolute error for model j
    e_i is the absolute value of the computed value minus the actual value for instance i
    n is the number of test set instances.

The hypothesis of no significant difference is then tested with the equation:

$$P = \frac{|mae_1 - mae_2|}{\sqrt{v(2/n)}} \tag{B.11}$$

where
    v is either the average or the larger of the variance scores for each model
    n is the total number of test set instances.

As is the case when the output attribute is categorical, using the larger of the two variance scores is the stronger test.

Subject Index

Index of R Functions

Script Index

9 780367 439149